Grades 1-2

BUILDING MATH SKILLS

BY Brian Rhee

Over 3600 Practice Problems

Detailed Solutions

Legal Notice

Copyright © 2018 by Solomon Academy
Published by: Solomon Academy
First Edition
ISBN-13: 978-1717048554
ISBN-10: 1717048552

All rights reserved. This publication or any portion thereof may not be copied, replicated, distributed, or transmitted in any form or by any means whether electronically or mechanically whatsoever. It is illegal to produce derivative works from this publication, in whole or in part, without the prior written permission of the publisher and author.

About This Book

This book is designed to help students build basic arithmetic and math skills. There are resources pages in the beginning of the book that illustrate how to do each type of problems. It is strongly recommended to go over the resources pages before solving practice problems.

This book contains 12 lessons with detailed solutions. Each lesson has ten practice worksheets which provides challenges to improve and strengthen students' math skills. Completing 12 lessons enables students to build their confidence and master their math skills.

About Author

Brian(Yeon) Rhee obtained a Masters of Arts Degree in Statistics at Columbia University, NY. He served as the Mathematical Statistician at the Bureau of Labor Statistics, DC. He is the Head Academic Director at Solomon Academy due to his devotion to the community coupled with his passion for teaching. His mission is to help students of all confidence level excel in academia to build a strong foundation in character, knowledge, and wisdom. Now, Solomon academy is known as the best academy specialized in Math in Northern Virginia.

Brian Rhee has published more than ten books. The titles of his books are 7 full-length practice tests for the AP Calculus AB/BC Multiple choice sections, AP Calculus, SAT 1 Math, SAT 2 Math level 2, 12 full-length practice tests for the SAT 2 Math Level 2, SHSAT/TJHSST Math workbook, and IAAT (Iowa Algebra Aptitude Test) Volume 1 and 2, CogAT form 7 Level 8, and NNAT 2 Level B Grade 1. He's currently working on other math books which will be introduced in the near future.

Brian Rhee has more than twenty years of teaching experience in math. He has been one of the most popular tutors among TJHSST (Thomas Jefferson High School For Science and Technology) students. Currently, he is developing many online math courses with www.masterprep.net for AP Calculus AB and BC, SAT 2 Math level 2 test, and other various math subjects.

SOLOMON ACADEMY

Solomon Academy is a prestigious institution of learning with numerous qualified teachers of various fields of education. Our mission is to thoroughly teach students of all ages and confidence levels, elevate skills to the highest standard of education, and provide them with all the tools and materials to succeed.

5723 Centre Square Drive
Centreville, VA 20120
Tel: 703-988-0019

Email: solomonacademyva@gmail.com
info@solomonacademy.net

CLASSES OFFERED

MATHEMATICS	TESTING	ENGLISH
1st-6th grade math	CogAt	1st-6th Reading
Algebra 1, 2	IAAT and SOL 7	1st-6th Writing
Geometry	TJHSST Prep	Essay Writing
Pre-Calculus	SAT/ACT Prep	SAT Writing
AP Calculus AB BC	SAT 2 Subject Tests	
AP Statistics	MathCounts	
Multivariate Calculus	AMC 10/12	

LEARN FROM THE AUTHOR

Private sessions with Brian Rhee is also available on the following subjects: SAT Math, SAT 2 Subject Math Level 2, Pre-Calculus, AP Calculus AB/BC, AP Statistics, IB SL/HL, Multivariate Calculus, Linear Algebra, AMC 8/10/12, and AIME.

Feel free to contact me at solomonacademyva@gmail.com

Acknowledgements

I wish to acknowledge my deepest appreciation to my wife, Sookyung, who has continuously given me wholehearted support, encouragement, and love. Without you, I could not have completed this book.

Thank you to my sons, Joshua and Jason, who have given me big smiles and inspiration. I love you all.

Thank you to Mr. Kwon from www.Masterprep.net, who has given me opportunities to develop online math courses for various math subjects.

Contents

How to do addition and subtraction with or without regrouping?

- Addition without regrouping
  1. Add the ones in ones column.
  2. Add the tens in tens column.

$$\begin{array}{r} 23 \\ +\ 45 \\ \hline 68 \end{array}$$

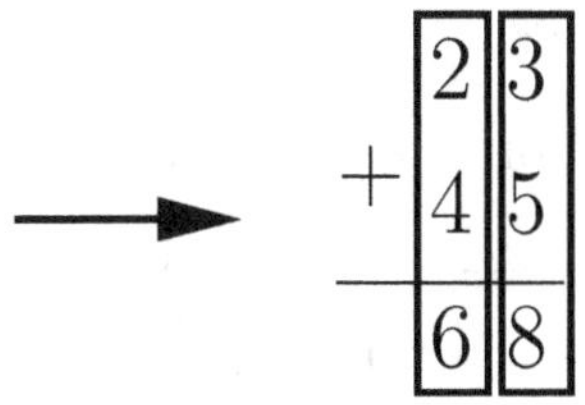

- Addition with regrouping
  1. Add the ones in ones column, then regroup.
  2. Add the tens in tens column, then regroup.

$$\begin{array}{r} 39 \\ +\ 24 \\ \hline 63 \end{array}$$

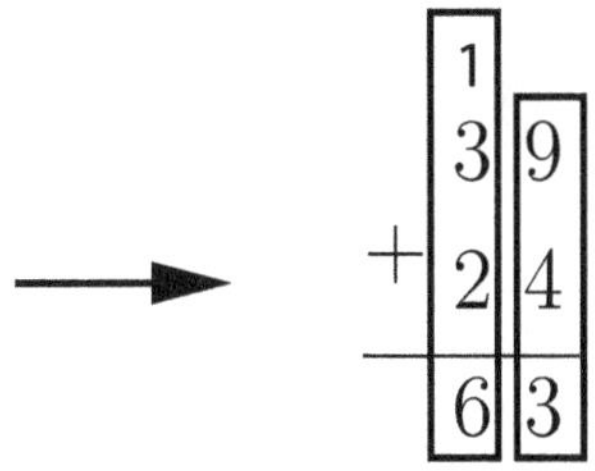

- Subtraction without regrouping
  1. Subtract the ones in ones column.

2. Subtract the tens in tens column.

$$\begin{array}{r} 7\,9 \\ -\;5\,4 \\ \hline 2\,5 \end{array} \longrightarrow$$

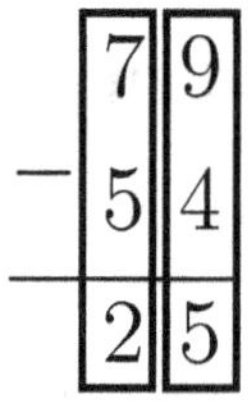

- Subtraction with regrouping

  1. Borrow 1 tens from the 4 in the tens place to subtract the ones in ones column.
  2. Change 2 as 12 and subtract the ones in ones column.
  3. Subtract the tens in tens column.

$$\begin{array}{r} 5\,2 \\ -\;1\,7 \\ \hline 3\,5 \end{array} \longrightarrow$$

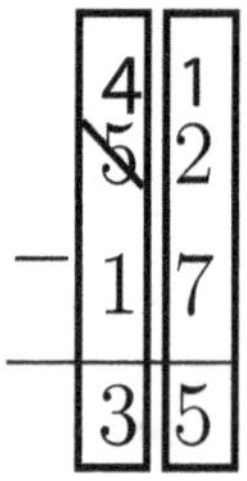

Name: **Lesson 1-1 Adding through 10**

1. 3 + 0	2. 1 + 4	3. 4 + 1	4. 0 + 2	5. 1 + 5
6. 2 + 3	7. 2 + 2	8. 0 + 5	9. 4 + 3	10. 8 + 0
11. 0 + 3	12. 2 + 5	13. 2 + 3	14. 1 + 6	15. 4 + 1
16. 6 + 3	17. 3 + 6	18. 5 + 4	19. 9 + 0	20. 2 + 6
21. 1 + 5	22. 5 + 0	23. 0 + 6	24. 2 + 7	25. 8 + 1
26. 3 + 3	27. 4 + 4	28. 1 + 7	29. 1 + 3	30. 4 + 2

Name: **Lesson 1-2 Adding through 11**

1. $\begin{array}{r} 1 \\ + \ 1 \\ \hline \end{array}$	2. $\begin{array}{r} 3 \\ + \ 2 \\ \hline \end{array}$	3. $\begin{array}{r} 3 \\ + \ 1 \\ \hline \end{array}$	4. $\begin{array}{r} 2 \\ + \ 3 \\ \hline \end{array}$	5. $\begin{array}{r} 3 \\ + \ 3 \\ \hline \end{array}$
6. $\begin{array}{r} 4 \\ + \ 1 \\ \hline \end{array}$	7. $\begin{array}{r} 1 \\ + \ 4 \\ \hline \end{array}$	8. $\begin{array}{r} 1 \\ + \ 3 \\ \hline \end{array}$	9. $\begin{array}{r} 6 \\ + \ 1 \\ \hline \end{array}$	10. $\begin{array}{r} 4 \\ + \ 5 \\ \hline \end{array}$
11. $\begin{array}{r} 2 \\ + \ 1 \\ \hline \end{array}$	12. $\begin{array}{r} 4 \\ + \ 3 \\ \hline \end{array}$	13. $\begin{array}{r} 0 \\ + \ 5 \\ \hline \end{array}$	14. $\begin{array}{r} 3 \\ + \ 4 \\ \hline \end{array}$	15. $\begin{array}{r} 2 \\ + \ 2 \\ \hline \end{array}$
16. $\begin{array}{r} 8 \\ + \ 1 \\ \hline \end{array}$	17. $\begin{array}{r} 5 \\ + \ 4 \\ \hline \end{array}$	18. $\begin{array}{r} 7 \\ + \ 2 \\ \hline \end{array}$	19. $\begin{array}{r} 2 \\ + \ 6 \\ \hline \end{array}$	20. $\begin{array}{r} 6 \\ + \ 2 \\ \hline \end{array}$
21. $\begin{array}{r} 3 \\ + \ 6 \\ \hline \end{array}$	22. $\begin{array}{r} 1 \\ + \ 7 \\ \hline \end{array}$	23. $\begin{array}{r} 2 \\ + \ 4 \\ \hline \end{array}$	24. $\begin{array}{r} 1 \\ + \ 6 \\ \hline \end{array}$	25. $\begin{array}{r} 2 \\ + \ 7 \\ \hline \end{array}$
26. $\begin{array}{r} 5 \\ + \ 6 \\ \hline \end{array}$	27. $\begin{array}{r} 7 \\ + \ 3 \\ \hline \end{array}$	28. $\begin{array}{r} 8 \\ + \ 2 \\ \hline \end{array}$	29. $\begin{array}{r} 6 \\ + \ 5 \\ \hline \end{array}$	30. $\begin{array}{r} 2 \\ + \ 9 \\ \hline \end{array}$

Name: **Lesson 1-3 Adding through 12**

1. $4 + 2$	2. $1 + 4$	3. $3 + 2$	4. $2 + 4$	5. $3 + 3$
6. $2 + 3$	7. $2 + 5$	8. $1 + 4$	9. $3 + 2$	10. $2 + 7$
11. $5 + 4$	12. $3 + 4$	13. $5 + 2$	14. $5 + 1$	15. $4 + 0$
16. $6 + 3$	17. $6 + 3$	18. $3 + 0$	19. $4 + 3$	20. $5 + 4$
21. $4 + 7$	22. $8 + 2$	23. $3 + 9$	24. $2 + 9$	25. $7 + 4$
26. $9 + 3$	27. $6 + 6$	28. $7 + 4$	29. $5 + 7$	30. $6 + 4$

Name: **Lesson 1-4 Adding through 13**

1. 5 + 2	2. 2 + 5	3. 5 + 3	4. 2 + 2	5. 3 + 1
6. 3 + 4	7. 8 + 0	8. 4 + 2	9. 3 + 5	10. 5 + 1
11. 4 + 4	12. 7 + 2	13. 4 + 3	14. 0 + 9	15. 2 + 6
16. 5 + 7	17. 2 + 4	18. 7 + 5	19. 9 + 2	20. 2 + 7
21. 8 + 4	22. 8 + 5	23. 9 + 4	24. 7 + 3	25. 9 + 3
26. 5 + 8	27. 4 + 9	28. 2 + 9	29. 3 + 9	30. 4 + 8

Name: **Lesson 1-5 Adding through 14**

1. $1 + 7$	2. $5 + 2$	3. $8 + 1$	4. $6 + 3$	5. $3 + 2$
6. $6 + 2$	7. $1 + 8$	8. $2 + 3$	9. $7 + 1$	10. $4 + 5$
11. $3 + 6$	12. $5 + 3$	13. $2 + 5$	14. $2 + 6$	15. $1 + 9$
16. $5 + 5$	17. $5 + 8$	18. $2 + 8$	19. $4 + 7$	20. $8 + 2$
21. $7 + 3$	22. $8 + 6$	23. $5 + 9$	24. $3 + 7$	25. $8 + 4$
26. $9 + 2$	27. $7 + 2$	28. $7 + 7$	29. $6 + 8$	30. $9 + 5$

Name: **Lesson 1-6 Adding through 15**

1. $3 + 4$	2. $6 + 1$	3. $3 + 2$	4. $5 + 4$	5. $4 + 4$
6. $2 + 7$	7. $2 + 6$	8. $5 + 6$	9. $7 + 3$	10. $7 + 0$
11. $3 + 5$	12. $4 + 6$	13. $8 + 5$	14. $6 + 5$	15. $6 + 6$
16. $6 + 4$	17. $5 + 7$	18. $7 + 7$	19. $7 + 8$	20. $5 + 8$
21. $8 + 7$	22. $9 + 6$	23. $5 + 9$	24. $8 + 7$	25. $7 + 5$
26. $6 + 9$	27. $6 + 7$	28. $7 + 8$	29. $7 + 6$	30. $9 + 5$

Name: **Lesson 1-7** **Adding through 16**

1. $\begin{array}{r} 3 \\ +\ 3 \\ \hline \end{array}$	2. $\begin{array}{r} 5 \\ +\ 3 \\ \hline \end{array}$	3. $\begin{array}{r} 2 \\ +\ 4 \\ \hline \end{array}$	4. $\begin{array}{r} 1 \\ +\ 7 \\ \hline \end{array}$	5. $\begin{array}{r} 3 \\ +\ 4 \\ \hline \end{array}$
6. $\begin{array}{r} 7 \\ +\ 2 \\ \hline \end{array}$	7. $\begin{array}{r} 2 \\ +\ 7 \\ \hline \end{array}$	8. $\begin{array}{r} 5 \\ +\ 2 \\ \hline \end{array}$	9. $\begin{array}{r} 0 \\ +\ 8 \\ \hline \end{array}$	10. $\begin{array}{r} 8 \\ +\ 1 \\ \hline \end{array}$
11. $\begin{array}{r} 4 \\ +\ 0 \\ \hline \end{array}$	12. $\begin{array}{r} 4 \\ +\ 6 \\ \hline \end{array}$	13. $\begin{array}{r} 9 \\ +\ 4 \\ \hline \end{array}$	14. $\begin{array}{r} 5 \\ +\ 6 \\ \hline \end{array}$	15. $\begin{array}{r} 7 \\ +\ 5 \\ \hline \end{array}$
16. $\begin{array}{r} 6 \\ +\ 5 \\ \hline \end{array}$	17. $\begin{array}{r} 6 \\ +\ 8 \\ \hline \end{array}$	18. $\begin{array}{r} 5 \\ +\ 7 \\ \hline \end{array}$	19. $\begin{array}{r} 5 \\ +\ 8 \\ \hline \end{array}$	20. $\begin{array}{r} 8 \\ +\ 6 \\ \hline \end{array}$
21. $\begin{array}{r} 7 \\ +\ 7 \\ \hline \end{array}$	22. $\begin{array}{r} 7 \\ +\ 6 \\ \hline \end{array}$	23. $\begin{array}{r} 7 \\ +\ 9 \\ \hline \end{array}$	24. $\begin{array}{r} 7 \\ +\ 8 \\ \hline \end{array}$	25. $\begin{array}{r} 5 \\ +\ 7 \\ \hline \end{array}$
26. $\begin{array}{r} 4 \\ +\ 9 \\ \hline \end{array}$	27. $\begin{array}{r} 8 \\ +\ 8 \\ \hline \end{array}$	28. $\begin{array}{r} 6 \\ +\ 7 \\ \hline \end{array}$	29. $\begin{array}{r} 9 \\ +\ 7 \\ \hline \end{array}$	30. $\begin{array}{r} 8 \\ +\ 7 \\ \hline \end{array}$

Name: **Lesson 1-8 Adding through 17**

1. $5 + 3$	2. $2 + 5$	3. $3 + 5$	4. $3 + 6$	5. $8 + 1$
6. $6 + 7$	7. $6 + 3$	8. $7 + 5$	9. $5 + 2$	10. $4 + 8$
11. $3 + 8$	12. $7 + 6$	13. $6 + 7$	14. $5 + 7$	15. $6 + 6$
16. $6 + 9$	17. $8 + 4$	18. $9 + 6$	19. $7 + 7$	20. $9 + 4$
21. $9 + 7$	22. $7 + 8$	23. $7 + 9$	24. $5 + 8$	25. $8 + 5$
26. $8 + 9$	27. $8 + 8$	28. $9 + 8$	29. $8 + 7$	30. $7 + 6$

Name: **Lesson 1-9 Adding through 18**

1. $3 + 7$	2. $4 + 6$	3. $8 + 2$	4. $1 + 9$	5. $9 + 1$
6. $6 + 4$	7. $5 + 7$	8. $6 + 4$	9. $7 + 4$	10. $5 + 6$
11. $7 + 6$	12. $4 + 7$	13. $7 + 5$	14. $5 + 8$	15. $6 + 8$
16. $6 + 7$	17. $8 + 7$	18. $8 + 6$	19. $7 + 8$	20. $7 + 7$
21. $9 + 7$	22. $9 + 6$	23. $5 + 9$	24. $9 + 8$	25. $9 + 5$
26. $9 + 9$	27. $8 + 9$	28. $8 + 8$	29. $7 + 9$	30. $6 + 9$

Name: **Lesson 1-10** **Adding through 19**

1. $2 + 9$	2. $5 + 5$	3. $9 + 2$	4. $6 + 5$	5. $1 + 9$
6. $3 + 10$	7. $5 + 6$	8. $6 + 8$	9. $8 + 6$	10. $6 + 6$
11. $3 + 8$	12. $6 + 9$	13. $7 + 7$	14. $8 + 7$	15. $7 + 8$
16. $4 + 7$	17. $7 + 10$	18. $5 + 8$	19. $9 + 6$	20. $8 + 5$
21. $8 + 8$	22. $9 + 7$	23. $10 + 8$	24. $10 + 7$	25. $9 + 8$
26. $10 + 9$	27. $8 + 9$	28. $9 + 10$	29. $7 + 9$	30. $8 + 10$

Name: **Lesson 2-1 Subtracting two numbers**

1. $3 - 1$	2. $4 - 2$	3. $4 - 1$	4. $5 - 3$	5. $3 - 0$
6. $5 - 2$	7. $5 - 4$	8. $5 - 1$	9. $6 - 4$	10. $5 - 5$
11. $6 - 3$	12. $6 - 2$	13. $7 - 4$	14. $7 - 3$	15. $6 - 5$
16. $7 - 2$	17. $7 - 1$	18. $8 - 1$	19. $8 - 5$	20. $7 - 6$
21. $9 - 5$	22. $8 - 7$	23. $8 - 8$	24. $9 - 3$	25. $8 - 4$
26. $8 - 6$	27. $9 - 4$	28. $9 - 8$	29. $9 - 9$	30. $9 - 2$

Name: **Lesson 2-2 Subtracting two numbers**

1. $6 - 4$	2. $9 - 8$	3. $6 - 5$	4. $9 - 4$	5. $8 - 7$
6. $7 - 6$	7. $5 - 2$	8. $5 - 1$	9. $7 - 5$	10. $8 - 4$
11. $8 - 1$	12. $7 - 2$	13. $8 - 0$	14. $9 - 3$	15. $9 - 7$
16. $6 - 6$	17. $9 - 4$	18. $5 - 3$	19. $5 - 4$	20. $9 - 2$
21. $5 - 5$	22. $8 - 2$	23. $8 - 7$	24. $6 - 3$	25. $7 - 6$
26. $7 - 0$	27. $5 - 3$	28. $6 - 2$	29. $7 - 4$	30. $9 - 3$

Name: **Lesson 2-3 Subtracting two numbers**

---

1. $8 - 1$	2. $5 - 3$	3. $6 - 2$	4. $5 - 2$	5. $8 - 3$
6. $4 - 0$	7. $3 - 1$	8. $4 - 2$	9. $8 - 7$	10. $6 - 5$
11. $7 - 3$	12. $8 - 6$	13. $7 - 4$	14. $9 - 9$	15. $7 - 0$
16. $5 - 4$	17. $5 - 5$	18. $6 - 3$	19. $3 - 2$	20. $7 - 2$
21. $6 - 0$	22. $7 - 5$	23. $8 - 5$	24. $5 - 1$	25. $4 - 4$
26. $7 - 7$	27. $6 - 1$	28. $7 - 1$	29. $8 - 2$	30. $6 - 0$

Name: Lesson 2-4 Subtracting two numbers

1. $3 - 0$	2. $8 - 2$	3. $5 - 1$	4. $6 - 3$	5. $8 - 5$
6. $7 - 4$	7. $2 - 0$	8. $4 - 3$	9. $7 - 0$	10. $3 - 2$
11. $5 - 3$	12. $6 - 0$	13. $2 - 1$	14. $8 - 4$	15. $6 - 2$
16. $7 - 6$	17. $7 - 1$	18. $8 - 6$	19. $4 - 0$	20. $7 - 5$
21. $4 - 1$	22. $4 - 4$	23. $8 - 3$	24. $8 - 1$	25. $5 - 4$
26. $6 - 4$	27. $7 - 2$	28. $7 - 3$	29. $8 - 7$	30. $8 - 8$

Name: **Lesson 2-5 Subtracting two numbers**

1. $\begin{array}{r} 9 \\ -\ 4 \\ \hline \end{array}$	2. $\begin{array}{r} 8 \\ -\ 2 \\ \hline \end{array}$	3. $\begin{array}{r} 6 \\ -\ 2 \\ \hline \end{array}$	4. $\begin{array}{r} 6 \\ -\ 3 \\ \hline \end{array}$	5. $\begin{array}{r} 9 \\ -\ 3 \\ \hline \end{array}$
6. $\begin{array}{r} 9 \\ -\ 0 \\ \hline \end{array}$	7. $\begin{array}{r} 7 \\ -\ 3 \\ \hline \end{array}$	8. $\begin{array}{r} 8 \\ -\ 4 \\ \hline \end{array}$	9. $\begin{array}{r} 8 \\ -\ 5 \\ \hline \end{array}$	10. $\begin{array}{r} 9 \\ -\ 8 \\ \hline \end{array}$
11. $\begin{array}{r} 3 \\ -\ 2 \\ \hline \end{array}$	12. $\begin{array}{r} 9 \\ -\ 9 \\ \hline \end{array}$	13. $\begin{array}{r} 6 \\ -\ 5 \\ \hline \end{array}$	14. $\begin{array}{r} 6 \\ -\ 3 \\ \hline \end{array}$	15. $\begin{array}{r} 8 \\ -\ 6 \\ \hline \end{array}$
16. $\begin{array}{r} 6 \\ -\ 4 \\ \hline \end{array}$	17. $\begin{array}{r} 7 \\ -\ 2 \\ \hline \end{array}$	18. $\begin{array}{r} 5 \\ -\ 2 \\ \hline \end{array}$	19. $\begin{array}{r} 8 \\ -\ 5 \\ \hline \end{array}$	20. $\begin{array}{r} 5 \\ -\ 5 \\ \hline \end{array}$
21. $\begin{array}{r} 8 \\ -\ 7 \\ \hline \end{array}$	22. $\begin{array}{r} 7 \\ -\ 7 \\ \hline \end{array}$	23. $\begin{array}{r} 7 \\ -\ 5 \\ \hline \end{array}$	24. $\begin{array}{r} 8 \\ -\ 3 \\ \hline \end{array}$	25. $\begin{array}{r} 4 \\ -\ 2 \\ \hline \end{array}$
26. $\begin{array}{r} 8 \\ -\ 0 \\ \hline \end{array}$	27. $\begin{array}{r} 5 \\ -\ 3 \\ \hline \end{array}$	28. $\begin{array}{r} 9 \\ -\ 5 \\ \hline \end{array}$	29. $\begin{array}{r} 4 \\ -\ 3 \\ \hline \end{array}$	30. $\begin{array}{r} 7 \\ -\ 6 \\ \hline \end{array}$

Name: **Lesson 2-6 Subtracting two numbers**

1. $8 - 6$	2. $5 - 2$	3. $6 - 4$	4. $7 - 5$	5. $3 - 3$
6. $4 - 2$	7. $8 - 8$	8. $4 - 0$	9. $6 - 4$	10. $9 - 8$
11. $9 - 9$	12. $5 - 0$	13. $9 - 7$	14. $8 - 5$	15. $3 - 2$
16. $5 - 4$	17. $8 - 3$	18. $8 - 2$	19. $7 - 3$	20. $7 - 4$
21. $9 - 4$	22. $4 - 3$	23. $7 - 6$	24. $9 - 2$	25. $5 - 3$
26. $9 - 0$	27. $9 - 3$	28. $9 - 5$	29. $9 - 6$	30. $6 - 2$

Name:

Lesson 2-7 Subtracting from 10, 11, and 12

1. $10 - 1$	2. $10 - 2$	3. $10 - 4$	4. $10 - 5$	5. $10 - 8$
6. $10 - 3$	7. $11 - 3$	8. $11 - 5$	9. $11 - 6$	10. $11 - 7$
11. $10 - 6$	12. $10 - 7$	13. $11 - 9$	14. $11 - 10$	15. $12 - 10$
16. $10 - 8$	17. $10 - 9$	18. $12 - 9$	19. $12 - 8$	20. $12 - 6$
21. $11 - 4$	22. $12 - 11$	23. $12 - 4$	24. $12 - 11$	25. $12 - 5$
26. $12 - 10$	27. $12 - 9$	28. $12 - 7$	29. $12 - 6$	30. $12 - 3$

Name:

Lesson 2-8 Subtracting from 13, 14, and 15

1. $\begin{array}{r} 13 \\ -\ 2 \\ \hline \end{array}$	2. $\begin{array}{r} 13 \\ -\ 4 \\ \hline \end{array}$	3. $\begin{array}{r} 13 \\ -\ 5 \\ \hline \end{array}$	4. $\begin{array}{r} 13 \\ -\ 9 \\ \hline \end{array}$	5. $\begin{array}{r} 13 \\ -\ 10 \\ \hline \end{array}$
6. $\begin{array}{r} 13 \\ -\ 0 \\ \hline \end{array}$	7. $\begin{array}{r} 13 \\ -\ 6 \\ \hline \end{array}$	8. $\begin{array}{r} 13 \\ -\ 8 \\ \hline \end{array}$	9. $\begin{array}{r} 14 \\ -\ 11 \\ \hline \end{array}$	10. $\begin{array}{r} 14 \\ -\ 4 \\ \hline \end{array}$
11. $\begin{array}{r} 14 \\ -\ 3 \\ \hline \end{array}$	12. $\begin{array}{r} 14 \\ -\ 0 \\ \hline \end{array}$	13. $\begin{array}{r} 14 \\ -\ 12 \\ \hline \end{array}$	14. $\begin{array}{r} 14 \\ -\ 8 \\ \hline \end{array}$	15. $\begin{array}{r} 14 \\ -\ 10 \\ \hline \end{array}$
16. $\begin{array}{r} 14 \\ -\ 14 \\ \hline \end{array}$	17. $\begin{array}{r} 15 \\ -\ 2 \\ \hline \end{array}$	18. $\begin{array}{r} 14 \\ -\ 9 \\ \hline \end{array}$	19. $\begin{array}{r} 15 \\ -\ 9 \\ \hline \end{array}$	20. $\begin{array}{r} 14 \\ -\ 7 \\ \hline \end{array}$
21. $\begin{array}{r} 15 \\ -\ 0 \\ \hline \end{array}$	22. $\begin{array}{r} 15 \\ -\ 14 \\ \hline \end{array}$	23. $\begin{array}{r} 15 \\ -\ 8 \\ \hline \end{array}$	24. $\begin{array}{r} 15 \\ -\ 6 \\ \hline \end{array}$	25. $\begin{array}{r} 15 \\ -\ 10 \\ \hline \end{array}$
26. $\begin{array}{r} 15 \\ -\ 13 \\ \hline \end{array}$	27. $\begin{array}{r} 15 \\ -\ 5 \\ \hline \end{array}$	28. $\begin{array}{r} 15 \\ -\ 7 \\ \hline \end{array}$	29. $\begin{array}{r} 15 \\ -\ 12 \\ \hline \end{array}$	30. $\begin{array}{r} 15 \\ -\ 4 \\ \hline \end{array}$

Name:

Lesson 2-9 Subtracting from 16, 17, and 18

1. $\begin{array}{r} 16 \\ -\ 1 \\ \hline \end{array}$

2. $\begin{array}{r} 16 \\ -\ 2 \\ \hline \end{array}$

3. $\begin{array}{r} 16 \\ -\ 3 \\ \hline \end{array}$

4. $\begin{array}{r} 16 \\ -\ 5 \\ \hline \end{array}$

5. $\begin{array}{r} 16 \\ -\ 15 \\ \hline \end{array}$

6. $\begin{array}{r} 16 \\ -\ 13 \\ \hline \end{array}$

7. $\begin{array}{r} 16 \\ -\ 11 \\ \hline \end{array}$

8. $\begin{array}{r} 16 \\ -\ 10 \\ \hline \end{array}$

9. $\begin{array}{r} 16 \\ -\ 9 \\ \hline \end{array}$

10. $\begin{array}{r} 16 \\ -\ 7 \\ \hline \end{array}$

11. $\begin{array}{r} 17 \\ -\ 17 \\ \hline \end{array}$

12. $\begin{array}{r} 16 \\ -\ 8 \\ \hline \end{array}$

13. $\begin{array}{r} 16 \\ -\ 7 \\ \hline \end{array}$

14. $\begin{array}{r} 16 \\ -\ 6 \\ \hline \end{array}$

15. $\begin{array}{r} 17 \\ -\ 3 \\ \hline \end{array}$

16. $\begin{array}{r} 17 \\ -\ 12 \\ \hline \end{array}$

17. $\begin{array}{r} 17 \\ -\ 5 \\ \hline \end{array}$

18. $\begin{array}{r} 17 \\ -\ 14 \\ \hline \end{array}$

19. $\begin{array}{r} 17 \\ -\ 11 \\ \hline \end{array}$

20. $\begin{array}{r} 17 \\ -\ 9 \\ \hline \end{array}$

21. $\begin{array}{r} 18 \\ -\ 18 \\ \hline \end{array}$

22. $\begin{array}{r} 17 \\ -\ 10 \\ \hline \end{array}$

23. $\begin{array}{r} 17 \\ -\ 8 \\ \hline \end{array}$

24. $\begin{array}{r} 18 \\ -\ 10 \\ \hline \end{array}$

25. $\begin{array}{r} 18 \\ -\ 12 \\ \hline \end{array}$

26. $\begin{array}{r} 18 \\ -\ 2 \\ \hline \end{array}$

27. $\begin{array}{r} 18 \\ -\ 9 \\ \hline \end{array}$

28. $\begin{array}{r} 18 \\ -\ 8 \\ \hline \end{array}$

29. $\begin{array}{r} 18 \\ -\ 7 \\ \hline \end{array}$

30. $\begin{array}{r} 18 \\ -\ 6 \\ \hline \end{array}$

Name: **Lesson 2-10 Subtracting from 17, 18, and 19**

1. $17 - 15$	2. $17 - 14$	3. $17 - 13$	4. $17 - 12$	5. $17 - 11$
6. $17 - 2$	7. $17 - 3$	8. $17 - 4$	9. $17 - 5$	10. $18 - 8$
11. $18 - 3$	12. $18 - 2$	13. $18 - 5$	14. $18 - 7$	15. $18 - 10$
16. $18 - 15$	17. $19 - 8$	18. $18 - 13$	19. $18 - 11$	20. $18 - 12$
21. $19 - 10$	22. $19 - 11$	23. $19 - 4$	24. $19 - 7$	25. $18 - 6$
26. $19 - 9$	27. $19 - 13$	28. $19 - 15$	29. $19 - 12$	30. $19 - 8$

Name: **Lesson 3-1** **Adding three numbers**

1. $3+6+1=$ 2. $5+5+2=$ 3. $1+2+3=$

4. $2+2+1=$ 5. $3+7+2=$ 6. $4+5+3=$

7. $4+5+1=$ 8. $7+4+2=$ 9. $1+4+3=$

10. $8+2+1=$ 11. $4+3+2=$ 12. $5+2+3=$

13. $6+4+1=$ 14. $2+6+2=$ 15. $2+7+3=$

16. $2+7+1=$ 17. $4+4+2=$ 18. $5+4+3=$

19. $5+4+1=$ 20. $2+8+2=$ 21. $6+1+3=$

22. $2+9+1=$ 23. $4+6+2=$ 24. $2+3+3=$

25. $3+2+1=$ 26. $5+3+2=$ 27. $5+1+3=$

28. $9+2+1=$ 29. $4+2+2=$ 30. $7+2+3=$

Name: **Lesson 3-2** **Adding three numbers**

1. $2+2+2=$	2. $3+6+3=$	3. $4+3+4=$
4. $6+4+2=$	5. $6+5+3=$	6. $2+4+4=$
7. $7+3+2=$	8. $8+2+3=$	9. $2+6+4=$
10. $7+2+2=$	11. $4+6+3=$	12. $5+5+4=$
13. $6+3+2=$	14. $2+8+3=$	15. $7+4+4=$
16. $8+3+2=$	17. $3+5+3=$	18. $4+4+4=$
19. $5+2+2=$	20. $5+4+3=$	21. $3+3+4=$
22. $3+8+2=$	23. $2+9+3=$	24. $2+3+4=$
25. $3+7+2=$	26. $4+7+3=$	27. $5+3+4=$
28. $5+6+2=$	29. $2+7+3=$	30. $4+5+4=$

1. $7+2+3=$	2. $5+5+4=$	3. $2+9+5=$
4. $3+5+3=$	5. $8+2+4=$	6. $4+5+5=$
7. $2+4+3=$	8. $3+6+4=$	9. $2+3+5=$
10. $4+3+3=$	11. $5+4+4=$	12. $5+2+5=$
13. $2+6+3=$	14. $8+3+4=$	15. $6+5+5=$
16. $4+7+3=$	17. $5+3+4=$	18. $2+2+5=$
19. $2+8+3=$	20. $7+3+4=$	21. $3+4+5=$
22. $4+6+3=$	23. $2+5+4=$	24. $3+8+5=$
25. $3+2+3=$	26. $6+4+4=$	27. $2+7+5=$
28. $6+3+3=$	29. $9+2+4=$	30. $4+4+5=$

Name: **Lesson 3-4** **Adding three numbers**

1. $8 + 2 + 4 =$
2. $7 + 2 + 5 =$
3. $3 + 5 + 6 =$
4. $3 + 2 + 4 =$
5. $2 + 6 + 5 =$
6. $5 + 2 + 6 =$
7. $9 + 2 + 4 =$
8. $3 + 4 + 5 =$
9. $2 + 2 + 6 =$
10. $7 + 4 + 4 =$
11. $5 + 6 + 5 =$
12. $4 + 2 + 6 =$
13. $5 + 4 + 4 =$
14. $2 + 3 + 5 =$
15. $6 + 3 + 6 =$
16. $2 + 8 + 4 =$
17. $3 + 7 + 5 =$
18. $2 + 9 + 6 =$
19. $2 + 5 + 4 =$
20. $8 + 3 + 5 =$
21. $6 + 4 + 6 =$
22. $7 + 3 + 4 =$
23. $4 + 5 + 5 =$
24. $5 + 5 + 6 =$
25. $6 + 5 + 4 =$
26. $3 + 8 + 5 =$
27. $2 + 4 + 6 =$
28. $6 + 3 + 4 =$
29. $2 + 7 + 5 =$
30. $4 + 7 + 6 =$

1. $9 + 2 + 5 =$	2. $6 + 3 + 6 =$	3. $5 + 2 + 7 =$
4. $5 + 3 + 5 =$	5. $4 + 7 + 6 =$	6. $3 + 8 + 7 =$
7. $2 + 6 + 5 =$	8. $2 + 3 + 6 =$	9. $2 + 8 + 7 =$
10. $5 + 4 + 5 =$	11. $4 + 2 + 6 =$	12. $3 + 3 + 7 =$
13. $2 + 7 + 5 =$	14. $3 + 4 + 6 =$	15. $7 + 3 + 7 =$
16. $3 + 5 + 5 =$	17. $4 + 6 + 6 =$	18. $2 + 4 + 7 =$
19. $4 + 4 + 5 =$	20. $2 + 9 + 6 =$	21. $6 + 4 + 7 =$
22. $5 + 1 + 5 =$	23. $3 + 7 + 6 =$	24. $7 + 2 + 7 =$
25. $3 + 2 + 5 =$	26. $4 + 5 + 6 =$	27. $2 + 5 + 7 =$
28. $8 + 2 + 5 =$	29. $8 + 3 + 6 =$	30. $7 + 4 + 7 =$

1. $4+3+6=$	2. $2+9+7=$	3. $4+2+8=$
4. $6+2+6=$	5. $3+3+7=$	6. $2+7+8=$
7. $5+4+6=$	8. $7+3+7=$	9. $4+4+8=$
10. $3+6+6=$	11. $2+2+7=$	12. $5+2+8=$
13. $6+5+6=$	14. $8+2+7=$	15. $4+7+8=$
16. $5+5+6=$	17. $6+3+7=$	18. $4+6+8=$
19. $8+3+6=$	20. $3+5+7=$	21. $2+4+8=$
22. $7+4+6=$	23. $6+4+7=$	24. $9+2+8=$
25. $3+2+6=$	26. $2+6+7=$	27. $3+8+8=$
28. $8+4+6=$	29. $3+7+7=$	30. $5+6+8=$

1. $9 + 2 + 7 =$
2. $5 + 4 + 8 =$
3. $2 + 2 + 9 =$
4. $2 + 6 + 7 =$
5. $8 + 2 + 8 =$
6. $4 + 3 + 9 =$
7. $3 + 3 + 7 =$
8. $2 + 3 + 8 =$
9. $6 + 2 + 9 =$
10. $2 + 5 + 7 =$
11. $7 + 2 + 8 =$
12. $5 + 6 + 9 =$
13. $7 + 4 + 7 =$
14. $2 + 4 + 8 =$
15. $8 + 3 + 9 =$
16. $2 + 7 + 7 =$
17. $3 + 8 + 8 =$
18. $4 + 4 + 9 =$
19. $6 + 3 + 7 =$
20. $6 + 4 + 8 =$
21. $3 + 6 + 9 =$
22. $5 + 3 + 7 =$
23. $5 + 5 + 8 =$
24. $2 + 8 + 9 =$
25. $3 + 7 + 7 =$
26. $3 + 2 + 8 =$
27. $4 + 6 + 9 =$
28. $5 + 2 + 7 =$
29. $4 + 5 + 8 =$
30. $7 + 3 + 9 =$

1. $5 + 2 + 8 =$	2. $7 + 4 + 9 =$	3. $2 + 6 + 10 =$
4. $6 + 5 + 8 =$	5. $3 + 5 + 9 =$	6. $4 + 5 + 10 =$
7. $7 + 2 + 8 =$	8. $6 + 3 + 9 =$	9. $5 + 5 + 10 =$
10. $4 + 3 + 8 =$	11. $2 + 4 + 9 =$	12. $3 + 2 + 10 =$
13. $8 + 2 + 8 =$	14. $4 + 6 + 9 =$	15. $9 + 2 + 10 =$
16. $3 + 7 + 8 =$	17. $3 + 8 + 9 =$	18. $4 + 7 + 10 =$
19. $6 + 4 + 8 =$	20. $3 + 4 + 9 =$	21. $2 + 5 + 10 =$
22. $5 + 4 + 8 =$	23. $2 + 2 + 9 =$	24. $7 + 3 + 10 =$
25. $3 + 6 + 8 =$	26. $5 + 3 + 9 =$	27. $5 + 6 + 10 =$
28. $4 + 4 + 8 =$	29. $2 + 9 + 9 =$	30. $6 + 2 + 10 =$

1. $2+4+1=$ 2. $5+6+1=$ 3. $3+8+1=$

4. $6+3+2=$ 5. $8+2+2=$ 6. $2+2+2=$

7. $4+6+3=$ 8. $9+2+3=$ 9. $4+3+3=$

10. $5+5+4=$ 11. $3+4+4=$ 12. $3+6+4=$

13. $3+4+5=$ 14. $2+5+5=$ 15. $2+3+5=$

16. $7+2+6=$ 17. $4+4+6=$ 18. $2+8+6=$

19. $2+6+7=$ 20. $2+4+7=$ 21. $4+7+7=$

22. $7+3+8=$ 23. $5+4+8=$ 24. $6+2+8=$

25. $6+5+9=$ 26. $3+9+9=$ 27. $2+7+9=$

28. $6+4+10=$ 29. $8+3+10=$ 30. $5+2+10=$

Name: **Lesson 3-10** **Adding three numbers**

1. $4 + 3 + 1 =$
2. $7 + 4 + 1 =$
3. $3 + 3 + 1 =$
4. $6 + 5 + 2 =$
5. $3 + 6 + 2 =$
6. $5 + 4 + 2 =$
7. $8 + 2 + 3 =$
8. $5 + 3 + 3 =$
9. $7 + 2 + 3 =$
10. $6 + 4 + 4 =$
11. $2 + 3 + 4 =$
12. $2 + 8 + 4 =$
13. $2 + 6 + 5 =$
14. $8 + 3 + 5 =$
15. $2 + 9 + 5 =$
16. $5 + 4 + 6 =$
17. $3 + 2 + 6 =$
18. $4 + 6 + 6 =$
19. $2 + 6 + 7 =$
20. $2 + 4 + 7 =$
21. $2 + 5 + 7 =$
22. $5 + 2 + 8 =$
23. $2 + 7 + 8 =$
24. $5 + 5 + 8 =$
25. $9 + 2 + 9 =$
26. $7 + 3 + 9 =$
27. $3 + 7 + 9 =$
28. $4 + 7 + 10 =$
29. $5 + 6 + 10 =$
30. $3 + 8 + 10 =$

Name:

Lesson 4-1 Adding and subtracting three numbers

1. $5 - 2 + 1 =$
2. $7 - 3 + 2 =$
3. $10 - 9 + 3 =$
4. $9 - 6 + 1 =$
5. $6 - 5 + 2 =$
6. $10 - 7 + 3 =$
7. $7 - 5 + 1 =$
8. $4 - 3 + 2 =$
9. $5 - 4 + 3 =$
10. $9 - 8 + 1 =$
11. $10 - 8 + 2 =$
12. $8 - 4 + 3 =$
13. $6 - 3 + 1 =$
14. $10 - 6 + 2 =$
15. $9 - 7 + 3 =$
16. $8 - 6 + 1 =$
17. $9 - 5 + 2 =$
18. $7 - 6 + 3 =$
19. $10 - 4 + 1 =$
20. $8 - 5 + 2 =$
21. $10 - 2 + 3 =$
22. $4 - 2 + 1 =$
23. $9 - 2 + 2 =$
24. $6 - 4 + 3 =$
25. $8 - 2 + 1 =$
26. $10 - 6 + 2 =$
27. $5 - 3 + 3 =$
28. $9 - 3 + 1 =$
29. $6 - 2 + 2 =$
30. $8 - 7 + 3 =$

Name: **Lesson 4-2 Adding and subtracting three numbers**

---

1. $9 - 3 + 2 =$
2. $7 - 4 + 3 =$
3. $10 - 4 + 4 =$
4. $6 - 3 + 2 =$
5. $5 - 2 + 3 =$
6. $9 - 7 + 4 =$
7. $10 - 8 + 2 =$
8. $7 - 6 + 3 =$
9. $10 - 9 + 4 =$
10. $8 - 4 + 2 =$
11. $4 - 3 + 3 =$
12. $9 - 5 + 4 =$
13. $9 - 7 + 2 =$
14. $8 - 5 + 3 =$
15. $6 - 5 + 4 =$
16. $5 - 4 + 2 =$
17. $5 - 3 + 3 =$
18. $10 - 7 + 4 =$
19. $7 - 2 + 2 =$
20. $9 - 2 + 3 =$
21. $3 - 2 + 4 =$
22. $10 - 3 + 2 =$
23. $6 - 4 + 3 =$
24. $8 - 2 + 4 =$
25. $6 - 2 + 2 =$
26. $7 - 5 + 3 =$
27. $4 - 2 + 4 =$
28. $10 - 5 + 2 =$
29. $9 - 6 + 3 =$
30. $10 - 6 + 4 =$

Name: **Lesson 4-3 Adding and subtracting three numbers**

1. $6-4+3=$ 2. $10-9+4=$ 3. $6-5+5=$

4. $9-4+3=$ 5. $10-8+4=$ 6. $5-2+5=$

7. $7-2+3=$ 8. $8-6+4=$ 9. $7-4+5=$

10. $10-6+3=$ 11. $9-8+4=$ 12. $10-4+5=$

13. $8-4+3=$ 14. $6-2+4=$ 15. $9-7+5=$

16. $3-2+3=$ 17. $7-5+4=$ 18. $10-3+5=$

19. $9-5+3=$ 20. $10-5+4=$ 21. $8-2+5=$

22. $5-4+3=$ 23. $4-2+4=$ 24. $10-7+5=$

25. $7-3+3=$ 26. $8-5+4=$ 27. $7-6+5=$

28. $10-2+3=$ 29. $9-3+4=$ 30. $9-6+5=$

Name: **Lesson 4-4 Adding and subtracting three numbers**

---

1. $10-3+4=$	2. $7-5+5=$	3. $8-4+6=$
4. $5-4+4=$	5. $10-5+5=$	6. $6-2+6=$
7. $3-2+4=$	8. $9-5+5=$	9. $5-3+6=$
10. $6-4+4=$	11. $10-9+5=$	12. $4-2+6=$
13. $8-7+4=$	14. $5-2+5=$	15. $9-6+6=$
16. $8-5+4=$	17. $7-3+5=$	18. $10-4+6=$
19. $9-2+4=$	20. $9-4+5=$	21. $7-6+6=$
22. $10-8+4=$	23. $7-4+5=$	24. $10-7+6=$
25. $4-3+4=$	26. $10-2+5=$	27. $8-6+6=$
28. $9-3+4=$	29. $8-3+5=$	30. $8-2+6=$

Name: **Lesson 4-5 Adding and subtracting three numbers**

---

1. $9 - 7 + 5 =$
2. $6 - 5 + 6 =$
3. $5 - 2 + 7 =$
4. $8 - 7 + 5 =$
5. $10 - 8 + 6 =$
6. $10 - 9 + 7 =$
7. $4 - 2 + 5 =$
8. $8 - 3 + 6 =$
9. $9 - 5 + 7 =$
10. $9 - 6 + 5 =$
11. $10 - 5 + 6 =$
12. $5 - 3 + 7 =$
13. $8 - 5 + 5 =$
14. $7 - 2 + 6 =$
15. $10 - 7 + 7 =$
16. $7 - 4 + 5 =$
17. $10 - 4 + 6 =$
18. $8 - 4 + 7 =$
19. $6 - 3 + 5 =$
20. $7 - 5 + 6 =$
21. $6 - 2 + 7 =$
22. $10 - 3 + 5 =$
23. $4 - 3 + 6 =$
24. $10 - 6 + 7 =$
25. $9 - 2 + 5 =$
26. $9 - 3 + 6 =$
27. $8 - 2 + 7 =$
28. $8 - 6 + 5 =$
29. $6 - 4 + 6 =$
30. $9 - 4 + 7 =$

Name: **Lesson 4-6 Adding and subtracting three numbers**

---

1. $9 - 4 + 6 =$ 2. $5 - 3 + 7 =$ 3. $9 - 8 + 8 =$

4. $10 - 5 + 6 =$ 5. $6 - 2 + 7 =$ 6. $8 - 6 + 8 =$

7. $3 - 2 + 6 =$ 8. $8 - 4 + 7 =$ 9. $10 - 4 + 8 =$

10. $6 - 4 + 6 =$ 11. $6 - 5 + 7 =$ 12. $4 - 2 + 8 =$

13. $8 - 7 + 6 =$ 14. $5 - 2 + 7 =$ 15. $7 - 6 + 8 =$

16. $9 - 7 + 6 =$ 17. $7 - 5 + 7 =$ 18. $10 - 6 + 8 =$

19. $10 - 7 + 6 =$ 20. $9 - 5 + 7 =$ 21. $5 - 4 + 8 =$

22. $8 - 2 + 6 =$ 23. $7 - 3 + 7 =$ 24. $9 - 6 + 8 =$

25. $7 - 3 + 6 =$ 26. $8 - 5 + 7 =$ 27. $9 - 2 + 8 =$

28. $10 - 9 + 6 =$ 29. $9 - 3 + 7 =$ 30. $7 - 4 + 8 =$

Name:

Lesson 4-7 Adding and subtracting three numbers

1. $5 - 4 + 7 =$
2. $7 - 4 + 8 =$
3. $8 - 4 + 9 =$
4. $8 - 2 + 7 =$
5. $10 - 7 + 8 =$
6. $6 - 3 + 9 =$
7. $10 - 8 + 7 =$
8. $9 - 8 + 8 =$
9. $5 - 2 + 9 =$
10. $6 - 5 + 7 =$
11. $9 - 7 + 8 =$
12. $7 - 5 + 9 =$
13. $8 - 3 + 7 =$
14. $10 - 2 + 8 =$
15. $5 - 3 + 9 =$
16. $7 - 3 + 7 =$
17. $4 - 3 + 8 =$
18. $10 - 4 + 9 =$
19. $6 - 2 + 7 =$
20. $8 - 6 + 8 =$
21. $10 - 9 + 9 =$
22. $10 - 5 + 7 =$
23. $9 - 4 + 8 =$
24. $7 - 2 + 9 =$
25. $8 - 7 + 7 =$
26. $10 - 6 + 8 =$
27. $6 - 4 + 9 =$
28. $9 - 3 + 7 =$
29. $9 - 2 + 8 =$
30. $7 - 6 + 9 =$

Name: **Lesson 4-8 Adding and subtracting three numbers**

1. $7 - 4 + 8 =$
2. $6 - 5 + 9 =$
3. $8 - 4 + 10 =$
4. $9 - 6 + 8 =$
5. $3 - 2 + 9 =$
6. $10 - 2 + 10 =$
7. $8 - 2 + 8 =$
8. $10 - 7 + 9 =$
9. $4 - 3 + 10 =$
10. $7 - 6 + 8 =$
11. $6 - 3 + 9 =$
12. $9 - 8 + 10 =$
13. $10 - 4 + 8 =$
14. $5 - 3 + 9 =$
15. $8 - 7 + 10 =$
16. $8 - 6 + 8 =$
17. $7 - 5 + 9 =$
18. $10 - 5 + 10 =$
19. $10 - 6 + 8 =$
20. $8 - 3 + 9 =$
21. $9 - 4 + 10 =$
22. $7 - 2 + 8 =$
23. $4 - 2 + 9 =$
24. $7 - 3 + 10 =$
25. $9 - 3 + 8 =$
26. $10 - 3 + 9 =$
27. $6 - 4 + 10 =$
28. $5 - 2 + 8 =$
29. $9 - 5 + 9 =$
30. $10 - 8 + 10 =$

Name:

Lesson 4-9 Adding and subtracting three numbers

1. $8 - 6 + 1 =$
2. $6 - 4 + 1 =$
3. $10 - 3 + 1 =$
4. $4 - 2 + 2 =$
5. $10 - 6 + 2 =$
6. $9 - 8 + 2 =$
7. $10 - 7 + 3 =$
8. $9 - 7 + 3 =$
9. $3 - 2 + 3 =$
10. $5 - 4 + 4 =$
11. $8 - 2 + 4 =$
12. $7 - 4 + 4 =$
13. $9 - 4 + 5 =$
14. $7 - 6 + 5 =$
15. $5 - 3 + 5 =$
16. $5 - 2 + 6 =$
17. $9 - 5 + 6 =$
18. $6 - 2 + 6 =$
19. $10 - 5 + 7 =$
20. $7 - 5 + 7 =$
21. $8 - 4 + 7 =$
22. $10 - 2 + 8 =$
23. $6 - 3 + 8 =$
24. $10 - 9 + 8 =$
25. $9 - 3 + 9 =$
26. $8 - 5 + 9 =$
27. $10 - 3 + 9 =$
28. $4 - 1 + 10 =$
29. $9 - 6 + 10 =$
30. $8 - 6 + 10 =$

Name:

Lesson 4-10 Adding and subtracting three numbers

1. $7 - 5 + 1 =$
2. $10 - 2 + 1 =$
3. $10 - 3 + 1 =$
4. $10 - 6 + 2 =$
5. $8 - 4 + 2 =$
6. $9 - 8 + 2 =$
7. $9 - 8 + 3 =$
8. $6 - 5 + 3 =$
9. $3 - 2 + 3 =$
10. $8 - 6 + 4 =$
11. $9 - 7 + 4 =$
12. $7 - 4 + 4 =$
13. $4 - 2 + 5 =$
14. $10 - 4 + 5 =$
15. $5 - 3 + 5 =$
16. $6 - 4 + 6 =$
17. $7 - 6 + 6 =$
18. $6 - 2 + 6 =$
19. $8 - 3 + 7 =$
20. $5 - 3 + 7 =$
21. $8 - 4 + 7 =$
22. $5 - 4 + 8 =$
23. $3 - 2 + 8 =$
24. $10 - 9 + 8 =$
25. $9 - 6 + 9 =$
26. $10 - 9 + 9 =$
27. $10 - 3 + 9 =$
28. $8 - 2 + 10 =$
29. $7 - 2 + 10 =$
30. $8 - 6 + 10 =$

Name: **Lesson 5-1 Adding one and two digit numbers**

1. $11 + 9$	2. $56 + 2$	3. $35 + 4$	4. $23 + 5$	5. $41 + 8$
6. $28 + 2$	7. $15 + 7$	8. $67 + 3$	9. $33 + 8$	10. $27 + 5$
11. $39 + 8$	12. $54 + 8$	13. $72 + 9$	14. $16 + 5$	15. $47 + 7$
16. $43 + 6$	17. $18 + 9$	18. $36 + 7$	19. $22 + 9$	20. $44 + 9$
21. $54 + 9$	22. $48 + 3$	23. $66 + 8$	24. $34 + 7$	25. $59 + 5$
26. $68 + 7$	27. $67 + 5$	28. $47 + 5$	29. $56 + 6$	30. $29 + 9$

Name:

Lesson 5-2 Adding one and two digit numbers

1. $\begin{array}{r} 25 \\ +\quad 3 \\ \hline \end{array}$
2. $\begin{array}{r} 52 \\ +\quad 6 \\ \hline \end{array}$
3. $\begin{array}{r} 18 \\ +\quad 2 \\ \hline \end{array}$
4. $\begin{array}{r} 37 \\ +\quad 2 \\ \hline \end{array}$
5. $\begin{array}{r} 46 \\ +\quad 7 \\ \hline \end{array}$
6. $\begin{array}{r} 68 \\ +\quad 5 \\ \hline \end{array}$
7. $\begin{array}{r} 17 \\ +\quad 4 \\ \hline \end{array}$
8. $\begin{array}{r} 78 \\ +\quad 5 \\ \hline \end{array}$
9. $\begin{array}{r} 25 \\ +\quad 9 \\ \hline \end{array}$
10. $\begin{array}{r} 58 \\ +\quad 4 \\ \hline \end{array}$
11. $\begin{array}{r} 37 \\ +\quad 6 \\ \hline \end{array}$
12. $\begin{array}{r} 12 \\ +\quad 9 \\ \hline \end{array}$
13. $\begin{array}{r} 48 \\ +\quad 3 \\ \hline \end{array}$
14. $\begin{array}{r} 64 \\ +\quad 6 \\ \hline \end{array}$
15. $\begin{array}{r} 17 \\ +\quad 3 \\ \hline \end{array}$
16. $\begin{array}{r} 76 \\ +\quad 4 \\ \hline \end{array}$
17. $\begin{array}{r} 36 \\ +\quad 8 \\ \hline \end{array}$
18. $\begin{array}{r} 35 \\ +\quad 8 \\ \hline \end{array}$
19. $\begin{array}{r} 13 \\ +\quad 8 \\ \hline \end{array}$
20. $\begin{array}{r} 28 \\ +\quad 5 \\ \hline \end{array}$
21. $\begin{array}{r} 65 \\ +\quad 6 \\ \hline \end{array}$
22. $\begin{array}{r} 67 \\ +\quad 4 \\ \hline \end{array}$
23. $\begin{array}{r} 65 \\ +\quad 6 \\ \hline \end{array}$
24. $\begin{array}{r} 45 \\ +\quad 6 \\ \hline \end{array}$
25. $\begin{array}{r} 37 \\ +\quad 7 \\ \hline \end{array}$
26. $\begin{array}{r} 27 \\ +\quad 9 \\ \hline \end{array}$
27. $\begin{array}{r} 38 \\ +\quad 5 \\ \hline \end{array}$
28. $\begin{array}{r} 59 \\ +\quad 9 \\ \hline \end{array}$
29. $\begin{array}{r} 74 \\ +\quad 7 \\ \hline \end{array}$
30. $\begin{array}{r} 15 \\ +\quad 9 \\ \hline \end{array}$

Name:

Lesson 5-3 Adding one and two digit numbers

1. $\begin{array}{r} 14 \\ +\ \ 4 \\ \hline \end{array}$ 2. $\begin{array}{r} 65 \\ +\ \ 3 \\ \hline \end{array}$ 3. $\begin{array}{r} 11 \\ +\ \ 7 \\ \hline \end{array}$ 4. $\begin{array}{r} 46 \\ +\ \ 2 \\ \hline \end{array}$ 5. $\begin{array}{r} 71 \\ +\ \ 8 \\ \hline \end{array}$

6. $\begin{array}{r} 36 \\ +\ \ 6 \\ \hline \end{array}$ 7. $\begin{array}{r} 78 \\ +\ \ 8 \\ \hline \end{array}$ 8. $\begin{array}{r} 32 \\ +\ \ 8 \\ \hline \end{array}$ 9. $\begin{array}{r} 18 \\ +\ \ 3 \\ \hline \end{array}$ 10. $\begin{array}{r} 28 \\ +\ \ 3 \\ \hline \end{array}$

11. $\begin{array}{r} 37 \\ +\ \ 5 \\ \hline \end{array}$ 12. $\begin{array}{r} 76 \\ +\ \ 8 \\ \hline \end{array}$ 13. $\begin{array}{r} 48 \\ +\ \ 8 \\ \hline \end{array}$ 14. $\begin{array}{r} 53 \\ +\ \ 7 \\ \hline \end{array}$ 15. $\begin{array}{r} 16 \\ +\ \ 4 \\ \hline \end{array}$

16. $\begin{array}{r} 64 \\ +\ \ 8 \\ \hline \end{array}$ 17. $\begin{array}{r} 15 \\ +\ \ 9 \\ \hline \end{array}$ 18. $\begin{array}{r} 15 \\ +\ \ 6 \\ \hline \end{array}$ 19. $\begin{array}{r} 27 \\ +\ \ 6 \\ \hline \end{array}$ 20. $\begin{array}{r} 44 \\ +\ \ 8 \\ \hline \end{array}$

21. $\begin{array}{r} 72 \\ +\ \ 9 \\ \hline \end{array}$ 22. $\begin{array}{r} 58 \\ +\ \ 4 \\ \hline \end{array}$ 23. $\begin{array}{r} 76 \\ +\ \ 7 \\ \hline \end{array}$ 24. $\begin{array}{r} 38 \\ +\ \ 7 \\ \hline \end{array}$ 25. $\begin{array}{r} 58 \\ +\ \ 5 \\ \hline \end{array}$

26. $\begin{array}{r} 28 \\ +\ \ 7 \\ \hline \end{array}$ 27. $\begin{array}{r} 69 \\ +\ \ 7 \\ \hline \end{array}$ 28. $\begin{array}{r} 38 \\ +\ \ 6 \\ \hline \end{array}$ 29. $\begin{array}{r} 29 \\ +\ \ 8 \\ \hline \end{array}$ 30. $\begin{array}{r} 28 \\ +\ \ 2 \\ \hline \end{array}$

Name:

Lesson 5-4 Adding one and two digit numbers

1. $17 + 1 =$
2. $30 + 8 =$
3. $76 + 2 =$
4. $23 + 5 =$
5. $58 + 2 =$
6. $13 + 7 =$
7. $68 + 8 =$
8. $45 + 8 =$
9. $18 + 7 =$
10. $36 + 4 =$
11. $78 + 4 =$
12. $28 + 4 =$
13. $15 + 5 =$
14. $64 + 7 =$
15. $47 + 7 =$
16. $55 + 6 =$
17. $16 + 5 =$
18. $37 + 4 =$
19. $76 + 8 =$
20. $12 + 9 =$
21. $48 + 8 =$
22. $67 + 6 =$
23. $54 + 6 =$
24. $26 + 5 =$
25. $58 + 5 =$
26. $25 + 5 =$
27. $56 + 9 =$
28. $43 + 8 =$
29. $36 + 8 =$
30. $17 + 6 =$

Name: **Lesson 5-5 Adding one and two digit numbers**

1. $\begin{array}{r} 26 \\ +\ 2 \\ \hline \end{array}$ 2. $\begin{array}{r} 64 \\ +\ 4 \\ \hline \end{array}$ 3. $\begin{array}{r} 71 \\ +\ 7 \\ \hline \end{array}$ 4. $\begin{array}{r} 37 \\ +\ 5 \\ \hline \end{array}$ 5. $\begin{array}{r} 18 \\ +\ 0 \\ \hline \end{array}$

6. $\begin{array}{r} 47 \\ +\ 7 \\ \hline \end{array}$ 7. $\begin{array}{r} 13 \\ +\ 9 \\ \hline \end{array}$ 8. $\begin{array}{r} 56 \\ +\ 4 \\ \hline \end{array}$ 9. $\begin{array}{r} 64 \\ +\ 7 \\ \hline \end{array}$ 10. $\begin{array}{r} 13 \\ +\ 7 \\ \hline \end{array}$

11. $\begin{array}{r} 78 \\ +\ 3 \\ \hline \end{array}$ 12. $\begin{array}{r} 18 \\ +\ 7 \\ \hline \end{array}$ 13. $\begin{array}{r} 48 \\ +\ 3 \\ \hline \end{array}$ 14. $\begin{array}{r} 56 \\ +\ 8 \\ \hline \end{array}$ 15. $\begin{array}{r} 24 \\ +\ 8 \\ \hline \end{array}$

16. $\begin{array}{r} 38 \\ +\ 2 \\ \hline \end{array}$ 17. $\begin{array}{r} 27 \\ +\ 8 \\ \hline \end{array}$ 18. $\begin{array}{r} 67 \\ +\ 6 \\ \hline \end{array}$ 19. $\begin{array}{r} 45 \\ +\ 5 \\ \hline \end{array}$ 20. $\begin{array}{r} 18 \\ +\ 5 \\ \hline \end{array}$

21. $\begin{array}{r} 54 \\ +\ 8 \\ \hline \end{array}$ 22. $\begin{array}{r} 75 \\ +\ 6 \\ \hline \end{array}$ 23. $\begin{array}{r} 38 \\ +\ 8 \\ \hline \end{array}$ 24. $\begin{array}{r} 62 \\ +\ 8 \\ \hline \end{array}$ 25. $\begin{array}{r} 17 \\ +\ 8 \\ \hline \end{array}$

26. $\begin{array}{r} 15 \\ +\ 8 \\ \hline \end{array}$ 27. $\begin{array}{r} 37 \\ +\ 5 \\ \hline \end{array}$ 28. $\begin{array}{r} 58 \\ +\ 4 \\ \hline \end{array}$ 29. $\begin{array}{r} 24 \\ +\ 6 \\ \hline \end{array}$ 30. $\begin{array}{r} 44 \\ +\ 9 \\ \hline \end{array}$

Name:

Lesson 5-6 Adding one and two digit numbers

1. 43 + 5 = ____
2. 11 + 8 = ____
3. 60 + 8 = ____
4. 34 + 4 = ____
5. 57 + 2 = ____
6. 78 + 2 = ____
7. 14 + 9 = ____
8. 23 + 7 = ____
9. 16 + 5 = ____
10. 38 + 6 = ____
11. 25 + 6 = ____
12. 41 + 9 = ____
13. 67 + 5 = ____
14. 58 + 5 = ____
15. 17 + 4 = ____
16. 76 + 8 = ____
17. 15 + 6 = ____
18. 38 + 7 = ____
19. 14 + 6 = ____
20. 52 + 8 = ____
21. 43 + 8 = ____
22. 67 + 6 = ____
23. 26 + 7 = ____
24. 74 + 7 = ____
25. 18 + 3 = ____
26. 32 + 9 = ____
27. 15 + 8 = ____
28. 75 + 5 = ____
29. 46 + 8 = ____
30. 67 + 7 = ____

Name: **Lesson 5-7 Adding one and two digit numbers**

1. $65 + 3$	2. $42 + 6$	3. $18 + 1$	4. $37 + 2$	5. $16 + 3$
6. $53 + 7$	7. $72 + 9$	8. $25 + 7$	9. $14 + 6$	10. $38 + 4$
11. $57 + 8$	12. $16 + 8$	13. $46 + 5$	14. $25 + 6$	15. $63 + 8$
16. $78 + 8$	17. $17 + 3$	18. $35 + 8$	19. $17 + 5$	20. $41 + 9$
21. $68 + 3$	22. $77 + 6$	23. $26 + 6$	24. $56 + 7$	25. $15 + 5$
26. $27 + 4$	27. $36 + 4$	28. $65 + 6$	29. $14 + 9$	30. $78 + 6$

Name: **Lesson 5-8 Adding one and two digit numbers**

1. $\begin{array}{r} 42 \\ +\ 6 \\ \hline \end{array}$
2. $\begin{array}{r} 64 \\ +\ 4 \\ \hline \end{array}$
3. $\begin{array}{r} 73 \\ +\ 5 \\ \hline \end{array}$
4. $\begin{array}{r} 27 \\ +\ 1 \\ \hline \end{array}$
5. $\begin{array}{r} 56 \\ +\ 4 \\ \hline \end{array}$
6. $\begin{array}{r} 38 \\ +\ 3 \\ \hline \end{array}$
7. $\begin{array}{r} 15 \\ +\ 9 \\ \hline \end{array}$
8. $\begin{array}{r} 17 \\ +\ 4 \\ \hline \end{array}$
9. $\begin{array}{r} 32 \\ +\ 9 \\ \hline \end{array}$
10. $\begin{array}{r} 17 \\ +\ 8 \\ \hline \end{array}$
11. $\begin{array}{r} 43 \\ +\ 7 \\ \hline \end{array}$
12. $\begin{array}{r} 28 \\ +\ 6 \\ \hline \end{array}$
13. $\begin{array}{r} 15 \\ +\ 7 \\ \hline \end{array}$
14. $\begin{array}{r} 54 \\ +\ 8 \\ \hline \end{array}$
15. $\begin{array}{r} 68 \\ +\ 5 \\ \hline \end{array}$
16. $\begin{array}{r} 74 \\ +\ 6 \\ \hline \end{array}$
17. $\begin{array}{r} 18 \\ +\ 8 \\ \hline \end{array}$
18. $\begin{array}{r} 38 \\ +\ 2 \\ \hline \end{array}$
19. $\begin{array}{r} 61 \\ +\ 9 \\ \hline \end{array}$
20. $\begin{array}{r} 24 \\ +\ 7 \\ \hline \end{array}$
21. $\begin{array}{r} 57 \\ +\ 6 \\ \hline \end{array}$
22. $\begin{array}{r} 45 \\ +\ 5 \\ \hline \end{array}$
23. $\begin{array}{r} 73 \\ +\ 7 \\ \hline \end{array}$
24. $\begin{array}{r} 16 \\ +\ 6 \\ \hline \end{array}$
25. $\begin{array}{r} 17 \\ +\ 5 \\ \hline \end{array}$
26. $\begin{array}{r} 43 \\ +\ 8 \\ \hline \end{array}$
27. $\begin{array}{r} 13 \\ +\ 8 \\ \hline \end{array}$
28. $\begin{array}{r} 57 \\ +\ 8 \\ \hline \end{array}$
29. $\begin{array}{r} 26 \\ +\ 7 \\ \hline \end{array}$
30. $\begin{array}{r} 36 \\ +\ 6 \\ \hline \end{array}$

Name:

Lesson 5-9 Adding one and two digit numbers

1. 48 + 1	2. 17 + 4	3. 62 + 6	4. 14 + 4	5. 32 + 8
6. 21 + 8	7. 76 + 6	8. 54 + 8	9. 28 + 3	10. 36 + 5
11. 12 + 9	12. 18 + 7	13. 65 + 6	14. 45 + 7	15. 53 + 8
16. 73 + 7	17. 15 + 5	18. 34 + 7	19. 18 + 6	20. 67 + 9
21. 23 + 8	22. 74 + 7	23. 46 + 6	24. 58 + 6	25. 77 + 8
26. 58 + 4	27. 25 + 9	28. 17 + 4	29. 18 + 4	30. 33 + 9

Name:

Lesson 5-10 Adding one and two digit numbers

1. $37 + 2$	2. $28 + 2$	3. $28 + 1$	4. $75 + 4$	5. $56 + 3$
6. $63 + 8$	7. $27 + 4$	8. $17 + 4$	9. $17 + 8$	10. $13 + 9$
11. $48 + 2$	12. $28 + 3$	13. $28 + 4$	14. $36 + 6$	15. $65 + 9$
16. $78 + 3$	17. $15 + 8$	18. $16 + 9$	19. $48 + 9$	20. $24 + 6$
21. $53 + 8$	22. $78 + 4$	23. $78 + 5$	24. $67 + 4$	25. $17 + 9$
26. $26 + 6$	27. $43 + 9$	28. $43 + 9$	29. $58 + 5$	30. $32 + 8$

Name: **Lesson 6-1 Subtracting one and two digit numbers**

1. $\begin{array}{r} 60 \\ -\ 8 \\ \hline \end{array}$	2. $\begin{array}{r} 21 \\ -\ 6 \\ \hline \end{array}$	3. $\begin{array}{r} 71 \\ -\ 4 \\ \hline \end{array}$	4. $\begin{array}{r} 45 \\ -\ 9 \\ \hline \end{array}$	5. $\begin{array}{r} 50 \\ -\ 7 \\ \hline \end{array}$
6. $\begin{array}{r} 39 \\ -\ 2 \\ \hline \end{array}$	7. $\begin{array}{r} 81 \\ -\ 5 \\ \hline \end{array}$	8. $\begin{array}{r} 13 \\ -\ 5 \\ \hline \end{array}$	9. $\begin{array}{r} 82 \\ -\ 8 \\ \hline \end{array}$	10. $\begin{array}{r} 42 \\ -\ 6 \\ \hline \end{array}$
11. $\begin{array}{r} 63 \\ -\ 8 \\ \hline \end{array}$	12. $\begin{array}{r} 51 \\ -\ 7 \\ \hline \end{array}$	13. $\begin{array}{r} 70 \\ -\ 7 \\ \hline \end{array}$	14. $\begin{array}{r} 31 \\ -\ 9 \\ \hline \end{array}$	15. $\begin{array}{r} 26 \\ -\ 8 \\ \hline \end{array}$
16. $\begin{array}{r} 12 \\ -\ 9 \\ \hline \end{array}$	17. $\begin{array}{r} 70 \\ -\ 3 \\ \hline \end{array}$	18. $\begin{array}{r} 87 \\ -\ 9 \\ \hline \end{array}$	19. $\begin{array}{r} 42 \\ -\ 5 \\ \hline \end{array}$	20. $\begin{array}{r} 62 \\ -\ 4 \\ \hline \end{array}$
21. $\begin{array}{r} 36 \\ -\ 9 \\ \hline \end{array}$	22. $\begin{array}{r} 11 \\ -\ 8 \\ \hline \end{array}$	23. $\begin{array}{r} 20 \\ -\ 5 \\ \hline \end{array}$	24. $\begin{array}{r} 50 \\ -\ 8 \\ \hline \end{array}$	25. $\begin{array}{r} 82 \\ -\ 7 \\ \hline \end{array}$
26. $\begin{array}{r} 65 \\ -\ 8 \\ \hline \end{array}$	27. $\begin{array}{r} 53 \\ -\ 7 \\ \hline \end{array}$	28. $\begin{array}{r} 43 \\ -\ 9 \\ \hline \end{array}$	29. $\begin{array}{r} 73 \\ -\ 6 \\ \hline \end{array}$	30. $\begin{array}{r} 31 \\ -\ 8 \\ \hline \end{array}$

Name:

Lesson 6-2 Subtracting one and two digit numbers

1. $11 - 7$	2. $31 - 8$	3. $61 - 6$	4. $27 - 9$	5. $84 - 6$
6. $71 - 9$	7. $42 - 4$	8. $51 - 3$	9. $12 - 5$	10. $21 - 9$
11. $32 - 6$	12. $72 - 7$	13. $60 - 5$	14. $81 - 8$	15. $40 - 4$
16. $72 - 8$	17. $20 - 9$	18. $51 - 7$	19. $63 - 6$	20. $26 - 8$
21. $45 - 7$	22. $32 - 9$	23. $71 - 3$	24. $81 - 5$	25. $11 - 4$
26. $76 - 9$	27. $24 - 6$	28. $83 - 8$	29. $50 - 3$	30. $60 - 6$

Name: **Lesson 6-3 Subtracting one and two digit numbers**

1. $74 - 9$	2. $42 - 8$	3. $63 - 6$	4. $51 - 7$	5. $12 - 9$
6. $23 - 8$	7. $82 - 7$	8. $31 - 4$	9. $35 - 9$	10. $41 - 3$
11. $51 - 2$	12. $21 - 9$	13. $14 - 6$	14. $73 - 7$	15. $60 - 5$
16. $82 - 6$	17. $14 - 7$	18. $34 - 8$	19. $50 - 4$	20. $26 - 9$
21. $41 - 8$	22. $60 - 7$	23. $72 - 4$	24. $81 - 9$	25. $51 - 6$
26. $11 - 4$	27. $31 - 5$	28. $65 - 7$	29. $20 - 5$	30. $75 - 8$

Name: **Lesson 6-4 Subtracting one and two digit numbers**

1. $54 - 6$	2. $30 - 3$	3. $43 - 7$	4. $72 - 9$	5. $31 - 7$
6. $31 - 8$	7. $61 - 9$	8. $81 - 5$	9. $53 - 4$	10. $64 - 6$
11. $16 - 7$	12. $90 - 6$	13. $21 - 9$	14. $82 - 8$	15. $31 - 5$
16. $35 - 7$	17. $23 - 6$	18. $44 - 9$	19. $33 - 5$	20. $50 - 8$
21. $30 - 9$	22. $71 - 7$	23. $63 - 8$	24. $80 - 6$	25. $61 - 7$
26. $54 - 9$	27. $22 - 4$	28. $31 - 4$	29. $27 - 9$	30. $81 - 8$

Name:

Lesson 6-5 Subtracting one and two digit numbers

1. 13 − 9	2. 51 − 8	3. 21 − 3	4. 84 − 9	5. 72 − 8
6. 30 − 8	7. 70 − 7	8. 42 − 4	9. 21 − 3	10. 71 − 8
11. 41 − 9	12. 24 − 6	13. 61 − 5	14. 50 − 7	15. 30 − 6
16. 82 − 7	17. 32 − 8	18. 21 − 3	19. 42 − 9	20. 21 − 5
21. 52 − 5	22. 80 − 4	23. 71 − 6	24. 61 − 4	25. 15 − 7
26. 21 − 6	27. 83 − 7	28. 43 − 8	29. 56 − 8	30. 63 − 6

Name:

Lesson 6-6 Subtracting one and two digit numbers

1. $63 - 8$	2. $43 - 9$	3. $72 - 7$	4. $21 - 5$	5. $32 - 6$
6. $22 - 5$	7. $80 - 6$	8. $53 - 8$	9. $41 - 4$	10. $71 - 5$
11. $14 - 9$	12. $39 - 5$	13. $83 - 7$	14. $20 - 2$	15. $61 - 4$
16. $31 - 9$	17. $14 - 5$	18. $60 - 8$	19. $31 - 3$	20. $42 - 6$
21. $54 - 8$	22. $74 - 7$	23. $21 - 3$	24. $80 - 9$	25. $51 - 7$
26. $40 - 3$	27. $23 - 6$	28. $65 - 9$	29. $37 - 8$	30. $33 - 9$

Name:

Lesson 6-7 Subtracting one and two digit numbers

1. $24 - 7$	2. $83 - 8$	3. $44 - 9$	4. $72 - 7$	5. $65 - 6$
6. $32 - 8$	7. $12 - 5$	8. $52 - 6$	9. $52 - 4$	10. $73 - 5$
11. $15 - 9$	12. $31 - 3$	13. $23 - 7$	14. $64 - 8$	15. $50 - 3$
16. $80 - 8$	17. $11 - 2$	18. $43 - 9$	19. $26 - 9$	20. $31 - 8$
21. $51 - 9$	22. $71 - 7$	23. $81 - 8$	24. $61 - 5$	25. $32 - 6$
26. $30 - 6$	27. $52 - 3$	28. $86 - 9$	29. $40 - 9$	30. $83 - 7$

Name: **Lesson 6-8 Subtracting one and two digit numbers**

1. $51 - 6$	2. $16 - 9$	3. $16 - 9$	4. $34 - 8$	5. $72 - 7$
6. $64 - 9$	7. $40 - 2$	8. $30 - 7$	9. $85 - 9$	10. $51 - 8$
11. $21 - 4$	12. $66 - 8$	13. $66 - 9$	14. $16 - 7$	15. $31 - 6$
16. $87 - 9$	17. $35 - 7$	18. $35 - 7$	19. $53 - 5$	20. $23 - 5$
21. $40 - 9$	22. $22 - 3$	23. $23 - 4$	24. $71 - 8$	25. $81 - 4$
26. $20 - 4$	27. $12 - 8$	28. $21 - 5$	29. $43 - 5$	30. $62 - 6$

Name:

Lesson 6-9 Subtracting one and two digit numbers

1. $15 - 6$	2. $34 - 9$	3. $62 - 5$	4. $61 - 6$	5. $94 - 8$
6. $43 - 8$	7. $71 - 5$	8. $23 - 8$	9. $32 - 5$	10. $74 - 5$
11. $70 - 5$	12. $21 - 7$	13. $64 - 4$	14. $91 - 9$	15. $42 - 8$
16. $41 - 9$	17. $80 - 5$	18. $82 - 7$	19. $53 - 5$	20. $35 - 5$
21. $34 - 6$	22. $34 - 5$	23. $24 - 6$	24. $90 - 3$	25. $51 - 3$
26. $36 - 9$	27. $65 - 8$	28. $41 - 8$	29. $65 - 8$	30. $93 - 5$

Name:

Lesson 6-10 Subtracting one and two digit numbers

1. 75 − 8	2. 41 − 4	3. 63 − 9	4. 91 − 9	5. 86 − 8
6. 37 − 9	7. 23 − 5	8. 51 − 8	9. 70 − 5	10. 61 − 7
11. 23 − 7	12. 46 − 7	13. 92 − 5	14. 51 − 7	15. 82 − 5
16. 32 − 6	17. 57 − 9	18. 87 − 9	19. 32 − 8	20. 43 − 7
21. 60 − 8	22. 90 − 4	23. 24 − 7	24. 74 − 5	25. 61 − 2
26. 73 − 5	27. 42 − 7	28. 53 − 6	29. 70 − 3	30. 36 − 9

1. $14+7+1=$ 2. $23+9+2=$ 3. $16+9+3=$

4. $18+4+1=$ 5. $15+8+2=$ 6. $29+9+3=$

7. $27+5+1=$ 8. $16+7+2=$ 9. $19+7+3=$

10. $16+6+1=$ 11. $29+5+2=$ 12. $18+8+3=$

13. $25+9+1=$ 14. $18+6+2=$ 15. $27+8+3=$

16. $26+8+1=$ 17. $27+7+2=$ 18. $14+6+3=$

19. $22+7+1=$ 20. $15+9+2=$ 21. $24+7+3=$

22. $19+4+1=$ 23. $24+8+2=$ 24. $15+6+3=$

25. $27+5+1=$ 26. $19+6+2=$ 27. $28+3+3=$

28. $26+8+1=$ 29. $29+3+2=$ 30. $15+7+3=$

1. $28 + 5 + 2 =$
2. $27 + 3 + 3 =$
3. $22 + 8 + 4 =$
4. $14 + 5 + 2 =$
5. $18 + 8 + 3 =$
6. $18 + 3 + 4 =$
7. $18 + 7 + 2 =$
8. $19 + 5 + 3 =$
9. $15 + 7 + 4 =$
10. $29 + 3 + 2 =$
11. $24 + 7 + 3 =$
12. $27 + 4 + 4 =$
13. $19 + 9 + 2 =$
14. $19 + 3 + 3 =$
15. $16 + 6 + 4 =$
16. $15 + 4 + 2 =$
17. $28 + 7 + 3 =$
18. $25 + 6 + 4 =$
19. $17 + 7 + 2 =$
20. $15 + 5 + 3 =$
21. $19 + 7 + 4 =$
22. $13 + 7 + 2 =$
23. $26 + 7 + 3 =$
24. $13 + 8 + 4 =$
25. $29 + 2 + 2 =$
26. $19 + 8 + 3 =$
27. $15 + 7 + 4 =$
28. $16 + 5 + 2 =$
29. $27 + 5 + 3 =$
30. $29 + 2 + 4 =$

1. $27+7+3=$	2. $26+7+4=$	3. $38+8+5=$
4. $38+4+3=$	5. $35+6+4=$	6. $29+3+5=$
7. $29+3+3=$	8. $18+6+4=$	9. $26+9+5=$
10. $16+5+3=$	11. $15+7+4=$	12. $19+9+5=$
13. $39+7+3=$	14. $27+8+4=$	15. $23+8+5=$
16. $17+5+3=$	17. $32+7+4=$	18. $17+6+5=$
19. $26+4+3=$	20. $17+7+4=$	21. $33+8+5=$
22. $19+7+3=$	23. $28+5+4=$	24. $13+9+5=$
25. $33+5+3=$	26. $37+4+4=$	27. $24+8+5=$
28. $17+9+3=$	29. $19+3+4=$	30. $35+5+5=$

1. $18 + 3 + 4 =$	2. $29 + 8 + 5 =$	3. $26 + 9 + 6 =$
4. $29 + 5 + 4 =$	5. $27 + 8 + 5 =$	6. $15 + 7 + 6 =$
7. $17 + 4 + 4 =$	8. $38 + 3 + 5 =$	9. $25 + 8 + 6 =$
10. $35 + 6 + 4 =$	11. $34 + 5 + 5 =$	12. $32 + 8 + 6 =$
13. $16 + 8 + 4 =$	14. $18 + 7 + 5 =$	15. $28 + 4 + 6 =$
16. $19 + 9 + 4 =$	17. $21 + 3 + 5 =$	18. $33 + 5 + 6 =$
19. $26 + 8 + 4 =$	20. $27 + 6 + 5 =$	21. $17 + 7 + 6 =$
22. $17 + 9 + 4 =$	23. $37 + 7 + 5 =$	24. $29 + 2 + 6 =$
25. $24 + 5 + 4 =$	26. $19 + 7 + 5 =$	27. $18 + 6 + 6 =$
28. $38 + 3 + 4 =$	29. $22 + 6 + 5 =$	30. $34 + 8 + 6 =$

1. $28 + 7 + 5 =$ 2. $33 + 9 + 6 =$ 3. $26 + 8 + 7 =$

4. $34 + 9 + 5 =$ 5. $45 + 9 + 6 =$ 6. $39 + 9 + 7 =$

7. $38 + 8 + 5 =$ 8. $29 + 6 + 6 =$ 9. $25 + 6 + 7 =$

10. $29 + 5 + 5 =$ 11. $28 + 9 + 6 =$ 12. $47 + 6 + 7 =$

13. $42 + 9 + 5 =$ 14. $35 + 8 + 6 =$ 15. $36 + 7 + 7 =$

16. $25 + 7 + 5 =$ 17. $28 + 3 + 6 =$ 18. $27 + 3 + 7 =$

19. $38 + 4 + 5 =$ 20. $39 + 2 + 6 =$ 21. $39 + 4 + 7 =$

22. $37 + 6 + 5 =$ 23. $44 + 5 + 6 =$ 24. $48 + 3 + 7 =$

25. $46 + 6 + 5 =$ 26. $24 + 8 + 6 =$ 27. $27 + 5 + 7 =$

28. $27 + 4 + 5 =$ 29. $36 + 7 + 6 =$ 30. $48 + 6 + 7 =$

1. $46+8+6=$	2. $25+7+7=$	3. $48+3+8=$
4. $39+3+6=$	5. $39+7+7=$	6. $37+6+8=$
7. $36+5+6=$	8. $29+4+7=$	9. $34+7+8=$
10. $44+8+6=$	11. $42+9+7=$	12. $28+7+8=$
13. $17+6+6=$	14. $46+6+7=$	15. $23+9+8=$
16. $18+9+6=$	17. $35+9+7=$	18. $47+4+8=$
19. $37+8+6=$	20. $28+9+7=$	21. $39+6+8=$
22. $46+9+6=$	23. $48+5+7=$	24. $46+7+8=$
25. $23+8+6=$	26. $35+6+7=$	27. $39+5+8=$
28. $29+3+6=$	29. $27+9+7=$	30. $47+5+8=$

1. $38 + 9 + 7 =$
2. $49 + 7 + 8 =$
3. $42 + 9 + 9 =$
4. $49 + 3 + 7 =$
5. $55 + 7 + 8 =$
6. $58 + 4 + 9 =$
7. $48 + 3 + 7 =$
8. $46 + 9 + 8 =$
9. $49 + 4 + 9 =$
10. $59 + 5 + 7 =$
11. $38 + 5 + 8 =$
12. $57 + 7 + 9 =$
13. $36 + 8 + 7 =$
14. $53 + 9 + 8 =$
15. $35 + 9 + 9 =$
16. $49 + 9 + 7 =$
17. $46 + 7 + 8 =$
18. $49 + 2 + 9 =$
19. $55 + 6 + 7 =$
20. $57 + 9 + 8 =$
21. $59 + 6 + 9 =$
22. $44 + 7 + 7 =$
23. $34 + 8 + 8 =$
24. $47 + 8 + 9 =$
25. $58 + 7 + 7 =$
26. $46 + 6 + 8 =$
27. $37 + 9 + 9 =$
28. $33 + 8 + 7 =$
29. $57 + 5 + 8 =$
30. $36 + 8 + 9 =$

Name: **Lesson 7-8** **Adding three numbers**

1. $56 + 9 + 8 =$ 2. $37 + 8 + 9 =$ 3. $47 + 6 + 10 =$

4. $48 + 3 + 8 =$ 5. $48 + 8 + 9 =$ 6. $59 + 7 + 10 =$

7. $39 + 2 + 8 =$ 8. $45 + 6 + 9 =$ 9. $36 + 6 + 10 =$

10. $57 + 7 + 8 =$ 11. $58 + 9 + 9 =$ 12. $45 + 7 + 10 =$

13. $39 + 4 + 8 =$ 14. $49 + 3 + 9 =$ 15. $55 + 8 + 10 =$

16. $44 + 9 + 8 =$ 17. $36 + 8 + 9 =$ 18. $47 + 6 + 10 =$

19. $35 + 8 + 8 =$ 20. $59 + 9 + 9 =$ 21. $38 + 4 + 10 =$

22. $45 + 7 + 8 =$ 23. $36 + 5 + 9 =$ 24. $48 + 9 + 10 =$

25. $58 + 5 + 8 =$ 26. $33 + 8 + 9 =$ 27. $39 + 5 + 10 =$

28. $31 + 9 + 8 =$ 29. $46 + 9 + 9 =$ 30. $44 + 6 + 10 =$

Name: Lesson 7-9 Adding three numbers

1. $35 + 9 + 1 =$	2. $38 + 5 + 1 =$	3. $39 + 3 + 1 =$
4. $48 + 8 + 2 =$	5. $57 + 6 + 2 =$	6. $36 + 8 + 2 =$
7. $57 + 4 + 3 =$	8. $45 + 6 + 3 =$	9. $48 + 4 + 3 =$
10. $49 + 4 + 4 =$	11. $33 + 9 + 4 =$	12. $57 + 8 + 4 =$
13. $58 + 9 + 5 =$	14. $49 + 5 + 5 =$	15. $46 + 7 + 5 =$
16. $36 + 6 + 6 =$	17. $56 + 5 + 6 =$	18. $36 + 7 + 6 =$
19. $39 + 6 + 7 =$	20. $48 + 3 + 7 =$	21. $45 + 6 + 7 =$
22. $44 + 7 + 8 =$	23. $37 + 7 + 8 =$	24. $59 + 2 + 8 =$
25. $57 + 5 + 9 =$	26. $44 + 9 + 9 =$	27. $39 + 8 + 9 =$
28. $49 + 9 + 10 =$	29. $53 + 6 + 10 =$	30. $49 + 7 + 10 =$

1. $54 + 7 + 1 =$ 2. $48 + 8 + 1 =$ 3. $39 + 7 + 1 =$

4. $44 + 9 + 2 =$ 5. $58 + 3 + 2 =$ 6. $58 + 9 + 2 =$

7. $46 + 8 + 3 =$ 8. $39 + 5 + 3 =$ 9. $49 + 9 + 3 =$

10. $59 + 2 + 4 =$ 11. $59 + 4 + 4 =$ 12. $38 + 4 + 4 =$

13. $57 + 4 + 5 =$ 14. $36 + 6 + 5 =$ 15. $59 + 4 + 5 =$

16. $44 + 8 + 6 =$ 17. $45 + 6 + 6 =$ 18. $47 + 6 + 6 =$

19. $36 + 9 + 7 =$ 20. $37 + 8 + 7 =$ 21. $33 + 8 + 7 =$

22. $58 + 7 + 8 =$ 23. $53 + 9 + 8 =$ 24. $47 + 5 + 8 =$

25. $47 + 7 + 9 =$ 26. $35 + 3 + 9 =$ 27. $59 + 6 + 9 =$

28. $58 + 5 + 10 =$ 29. $46 + 5 + 10 =$ 30. $47 + 9 + 10 =$

Name:

Lesson 8-1 Adding and subtracting three numbers

1. $14 - 6 + 1 =$

2. $23 - 6 + 2 =$

3. $36 - 9 + 3 =$

4. $25 - 7 + 1 =$

5. $37 - 9 + 2 =$

6. $24 - 8 + 3 =$

7. $21 - 6 + 1 =$

8. $13 - 4 + 2 =$

9. $16 - 6 + 3 =$

10. $31 - 5 + 1 =$

11. $22 - 5 + 2 =$

12. $24 - 6 + 3 =$

13. $26 - 7 + 1 =$

14. $33 - 7 + 2 =$

15. $31 - 5 + 3 =$

16. $32 - 8 + 1 =$

17. $31 - 8 + 2 =$

18. $12 - 3 + 3 =$

19. $25 - 6 + 1 =$

20. $24 - 6 + 2 =$

21. $27 - 8 + 3 =$

22. $31 - 5 + 1 =$

23. $36 - 8 + 2 =$

24. $23 - 2 + 3 =$

25. $28 - 9 + 1 =$

26. $26 - 8 + 2 =$

27. $31 - 3 + 3 =$

28. $23 - 8 + 1 =$

29. $16 - 7 + 2 =$

30. $25 - 9 + 3 =$

Name:

Lesson 8-2 Adding and subtracting three numbers

1. $22 - 5 + 2 =$
2. $25 - 7 + 3 =$
3. $31 - 5 + 4 =$
4. $11 - 4 + 2 =$
5. $34 - 3 + 3 =$
6. $17 - 8 + 4 =$
7. $36 - 7 + 2 =$
8. $22 - 4 + 3 =$
9. $23 - 9 + 4 =$
10. $24 - 6 + 2 =$
11. $34 - 6 + 3 =$
12. $36 - 8 + 4 =$
13. $21 - 6 + 2 =$
14. $23 - 8 + 3 =$
15. $23 - 7 + 4 =$
16. $14 - 9 + 2 =$
17. $31 - 2 + 3 =$
18. $15 - 6 + 4 =$
19. $22 - 8 + 2 =$
20. $17 - 9 + 3 =$
21. $24 - 7 + 4 =$
22. $16 - 8 + 2 =$
23. $33 - 5 + 3 =$
24. $21 - 9 + 4 =$
25. $33 - 4 + 2 =$
26. $11 - 8 + 3 =$
27. $18 - 9 + 4 =$
28. $32 - 9 + 2 =$
29. $25 - 8 + 3 =$
30. $31 - 7 + 4 =$

Name: **Lesson 8-3 Adding and subtracting three numbers**

---

1. $25 - 8 + 3 =$
2. $31 - 9 + 4 =$
3. $36 - 8 + 5 =$
4. $31 - 4 + 3 =$
5. $27 - 8 + 4 =$
6. $32 - 9 + 5 =$
7. $32 - 6 + 3 =$
8. $33 - 6 + 4 =$
9. $21 - 7 + 5 =$
10. $47 - 9 + 3 =$
11. $21 - 2 + 4 =$
12. $33 - 4 + 5 =$
13. $32 - 7 + 3 =$
14. $33 - 8 + 4 =$
15. $42 - 5 + 5 =$
16. $44 - 9 + 3 =$
17. $44 - 7 + 4 =$
18. $22 - 8 + 5 =$
19. $36 - 7 + 3 =$
20. $31 - 8 + 4 =$
21. $34 - 8 + 5 =$
22. $25 - 7 + 3 =$
23. $42 - 3 + 4 =$
24. $38 - 9 + 5 =$
25. $31 - 6 + 3 =$
26. $35 - 9 + 4 =$
27. $26 - 7 + 5 =$
28. $21 - 5 + 3 =$
29. $24 - 6 + 4 =$
30. $41 - 3 + 5 =$

Name: **Lesson 8-4 Adding and subtracting three numbers**

---

1. $24 - 9 + 4 =$ 2. $21 - 6 + 5 =$ 3. $45 - 6 + 6 =$

4. $37 - 8 + 4 =$ 5. $36 - 8 + 5 =$ 6. $33 - 4 + 6 =$

7. $24 - 5 + 4 =$ 8. $32 - 8 + 5 =$ 9. $24 - 8 + 6 =$

10. $35 - 7 + 4 =$ 11. $43 - 7 + 5 =$ 12. $35 - 9 + 6 =$

13. $43 - 8 + 4 =$ 14. $32 - 5 + 5 =$ 15. $23 - 4 + 6 =$

16. $41 - 9 + 4 =$ 17. $28 - 9 + 5 =$ 18. $31 - 8 + 6 =$

19. $32 - 7 + 4 =$ 20. $23 - 6 + 5 =$ 21. $46 - 7 + 6 =$

22. $41 - 3 + 4 =$ 23. $42 - 5 + 5 =$ 24. $32 - 6 + 6 =$

25. $24 - 8 + 4 =$ 26. $35 - 8 + 5 =$ 27. $41 - 4 + 6 =$

28. $44 - 6 + 4 =$ 29. $31 - 6 + 5 =$ 30. $22 - 4 + 6 =$

Name: **Lesson 8-5 Adding and subtracting three numbers**

1. $33 - 5 + 5 =$	2. $34 - 7 + 6 =$	3. $35 - 6 + 7 =$
4. $42 - 6 + 5 =$	5. $43 - 4 + 6 =$	6. $47 - 8 + 7 =$
7. $52 - 9 + 5 =$	8. $46 - 9 + 6 =$	9. $51 - 5 + 7 =$
10. $34 - 8 + 5 =$	11. $54 - 6 + 6 =$	12. $44 - 9 + 7 =$
13. $41 - 8 + 5 =$	14. $32 - 3 + 6 =$	15. $46 - 8 + 7 =$
16. $54 - 7 + 5 =$	17. $43 - 9 + 6 =$	18. $36 - 7 + 7 =$
19. $45 - 8 + 5 =$	20. $31 - 2 + 6 =$	21. $55 - 9 + 7 =$
22. $55 - 7 + 5 =$	23. $51 - 6 + 6 =$	24. $52 - 4 + 7 =$
25. $41 - 8 + 5 =$	26. $45 - 8 + 6 =$	27. $43 - 7 + 7 =$
28. $32 - 9 + 5 =$	29. $44 - 5 + 6 =$	30. $52 - 8 + 7 =$

Name: **Lesson 8-6 Adding and subtracting three numbers**

1. $36 - 9 + 6 =$	2. $42 - 7 + 7 =$	3. $32 - 3 + 8 =$
4. $45 - 7 + 6 =$	5. $34 - 8 + 7 =$	6. $41 - 7 + 8 =$
7. $33 - 8 + 6 =$	8. $51 - 5 + 7 =$	9. $53 - 4 + 8 =$
10. $44 - 6 + 6 =$	11. $43 - 5 + 7 =$	12. $32 - 7 + 8 =$
13. $51 - 4 + 6 =$	14. $52 - 4 + 7 =$	15. $53 - 6 + 8 =$
16. $48 - 9 + 6 =$	17. $41 - 3 + 7 =$	18. $47 - 9 + 8 =$
19. $32 - 8 + 6 =$	20. $35 - 7 + 7 =$	21. $36 - 8 + 8 =$
22. $33 - 7 + 6 =$	23. $55 - 8 + 7 =$	24. $31 - 2 + 8 =$
25. $45 - 9 + 6 =$	26. $31 - 9 + 7 =$	27. $43 - 9 + 8 =$
28. $55 - 7 + 6 =$	29. $44 - 5 + 7 =$	30. $52 - 5 + 8 =$

Name: **Lesson 8-7 Adding and subtracting three numbers**

1. $45 - 7 + 7 =$
2. $65 - 5 + 8 =$
3. $43 - 8 + 9 =$
4. $58 - 9 + 7 =$
5. $54 - 6 + 8 =$
6. $61 - 9 + 9 =$
7. $43 - 8 + 7 =$
8. $56 - 9 + 8 =$
9. $56 - 8 + 9 =$
10. $44 - 6 + 7 =$
11. $43 - 5 + 8 =$
12. $44 - 8 + 9 =$
13. $51 - 2 + 7 =$
14. $61 - 4 + 8 =$
15. $57 - 8 + 9 =$
16. $42 - 8 + 7 =$
17. $52 - 8 + 8 =$
18. $41 - 6 + 9 =$
19. $62 - 6 + 7 =$
20. $61 - 3 + 8 =$
21. $46 - 7 + 9 =$
22. $52 - 4 + 7 =$
23. $57 - 9 + 8 =$
24. $52 - 5 + 9 =$
25. $43 - 9 + 7 =$
26. $42 - 4 + 8 =$
27. $64 - 7 + 9 =$
28. $51 - 8 + 7 =$
29. $51 - 5 + 8 =$
30. $53 - 6 + 9 =$

Name: **Lesson 8-8 Adding and subtracting three numbers**

---

1. $43 - 7 + 8 =$
2. $41 - 9 + 9 =$
3. $61 - 9 + 10 =$
4. $52 - 5 + 8 =$
5. $53 - 4 + 9 =$
6. $52 - 5 + 10 =$
7. $65 - 7 + 8 =$
8. $64 - 7 + 9 =$
9. $67 - 8 + 10 =$
10. $64 - 5 + 8 =$
11. $41 - 3 + 9 =$
12. $42 - 6 + 10 =$
13. $51 - 3 + 8 =$
14. $64 - 5 + 9 =$
15. $54 - 7 + 10 =$
16. $52 - 9 + 8 =$
17. $58 - 9 + 9 =$
18. $44 - 5 + 10 =$
19. $42 - 4 + 8 =$
20. $46 - 8 + 9 =$
21. $51 - 5 + 10 =$
22. $53 - 8 + 8 =$
23. $57 - 8 + 9 =$
24. $62 - 5 + 10 =$
25. $67 - 9 + 8 =$
26. $44 - 8 + 9 =$
27. $41 - 8 + 10 =$
28. $44 - 9 + 8 =$
29. $55 - 7 + 9 =$
30. $53 - 9 + 10 =$

Name: **Lesson 8-9 Adding and subtracting three numbers**

---

1. $42 - 4 + 1 =$ 2. $36 - 8 + 1 =$ 3. $61 - 9 + 1 =$

4. $35 - 8 + 2 =$ 5. $23 - 6 + 2 =$ 6. $22 - 3 + 2 =$

7. $62 - 9 + 3 =$ 8. $65 - 7 + 3 =$ 9. $41 - 6 + 3 =$

10. $55 - 9 + 4 =$ 11. $31 - 3 + 4 =$ 12. $57 - 9 + 4 =$

13. $41 - 8 + 5 =$ 14. $43 - 4 + 5 =$ 15. $34 - 7 + 5 =$

16. $24 - 8 + 6 =$ 17. $58 - 9 + 6 =$ 18. $24 - 5 + 6 =$

19. $33 - 6 + 7 =$ 20. $27 - 9 + 7 =$ 21. $51 - 5 + 7 =$

22. $63 - 7 + 8 =$ 23. $14 - 9 + 8 =$ 24. $63 - 7 + 8 =$

25. $52 - 5 + 9 =$ 26. $52 - 7 + 9 =$ 27. $33 - 6 + 9 =$

28. $21 - 4 + 10 =$ 29. $43 - 6 + 10 =$ 30. $56 - 7 + 10 =$

Name: **Lesson 8-10 Adding and subtracting three numbers**

1. $23 - 7 + 1 =$
2. $27 - 9 + 1 =$
3. $42 - 9 + 1 =$
4. $46 - 9 + 2 =$
5. $42 - 4 + 2 =$
6. $36 - 8 + 2 =$
7. $41 - 5 + 3 =$
8. $35 - 7 + 3 =$
9. $31 - 6 + 3 =$
10. $34 - 7 + 4 =$
11. $65 - 6 + 4 =$
12. $22 - 8 + 4 =$
13. $53 - 5 + 5 =$
14. $33 - 4 + 5 =$
15. $63 - 9 + 5 =$
16. $64 - 5 + 6 =$
17. $56 - 7 + 6 =$
18. $58 - 9 + 6 =$
19. $21 - 9 + 7 =$
20. $32 - 7 + 7 =$
21. $31 - 3 + 7 =$
22. $65 - 9 + 8 =$
23. $54 - 5 + 8 =$
24. $42 - 5 + 8 =$
25. $34 - 7 + 9 =$
26. $32 - 8 + 9 =$
27. $62 - 8 + 9 =$
28. $52 - 8 + 10 =$
29. $61 - 2 + 10 =$
30. $55 - 9 + 10 =$

Name: **Lesson 1 Adding two digit numbers**

1. $\begin{array}{r} 54 \\ +\ 20 \\ \hline \end{array}$	2. $\begin{array}{r} 35 \\ +\ 23 \\ \hline \end{array}$	3. $\begin{array}{r} 36 \\ +\ 52 \\ \hline \end{array}$	4. $\begin{array}{r} 86 \\ +\ 23 \\ \hline \end{array}$	5. $\begin{array}{r} 72 \\ +\ 13 \\ \hline \end{array}$
6. $\begin{array}{r} 27 \\ +\ 72 \\ \hline \end{array}$	7. $\begin{array}{r} 25 \\ +\ 41 \\ \hline \end{array}$	8. $\begin{array}{r} 32 \\ +\ 12 \\ \hline \end{array}$	9. $\begin{array}{r} 85 \\ +\ 16 \\ \hline \end{array}$	10. $\begin{array}{r} 46 \\ +\ 23 \\ \hline \end{array}$
11. $\begin{array}{r} 72 \\ +\ 30 \\ \hline \end{array}$	12. $\begin{array}{r} 53 \\ +\ 28 \\ \hline \end{array}$	13. $\begin{array}{r} 24 \\ +\ 18 \\ \hline \end{array}$	14. $\begin{array}{r} 39 \\ +\ 31 \\ \hline \end{array}$	15. $\begin{array}{r} 43 \\ +\ 39 \\ \hline \end{array}$
16. $\begin{array}{r} 35 \\ +\ 46 \\ \hline \end{array}$	17. $\begin{array}{r} 44 \\ +\ 36 \\ \hline \end{array}$	18. $\begin{array}{r} 66 \\ +\ 28 \\ \hline \end{array}$	19. $\begin{array}{r} 27 \\ +\ 66 \\ \hline \end{array}$	20. $\begin{array}{r} 29 \\ +\ 49 \\ \hline \end{array}$
21. $\begin{array}{r} 32 \\ +\ 95 \\ \hline \end{array}$	22. $\begin{array}{r} 96 \\ +\ 82 \\ \hline \end{array}$	23. $\begin{array}{r} 93 \\ +\ 24 \\ \hline \end{array}$	24. $\begin{array}{r} 87 \\ +\ 70 \\ \hline \end{array}$	25. $\begin{array}{r} 67 \\ +\ 53 \\ \hline \end{array}$
26. $\begin{array}{r} 85 \\ +\ 44 \\ \hline \end{array}$	27. $\begin{array}{r} 78 \\ +\ 61 \\ \hline \end{array}$	28. $\begin{array}{r} 42 \\ +\ 72 \\ \hline \end{array}$	29. $\begin{array}{r} 75 \\ +\ 82 \\ \hline \end{array}$	30. $\begin{array}{r} 53 \\ +\ 55 \\ \hline \end{array}$

Name: **Lesson 9-2 Adding two digit numbers**

1. $34 + 61$	2. $53 + 30$	3. $42 + 44$	4. $57 + 12$	5. $84 + 13$
6. $48 + 12$	7. $36 + 51$	8. $73 + 13$	9. $95 + 26$	10. $43 + 45$
11. $80 + 22$	12. $68 + 13$	13. $38 + 32$	14. $45 + 15$	15. $59 + 23$
16. $57 + 13$	17. $89 + 15$	18. $29 + 43$	19. $68 + 26$	20. $39 + 22$
21. $47 + 61$	22. $81 + 24$	23. $79 + 51$	24. $96 + 62$	25. $94 + 75$
26. $91 + 70$	27. $36 + 93$	28. $84 + 51$	29. $87 + 53$	30. $65 + 63$

Name: ______ **Lesson 9-3 Adding two digit numbers**

1. $\begin{array}{r} 54 \\ +\ 12 \\ \hline \end{array}$ 2. $\begin{array}{r} 26 \\ +\ 41 \\ \hline \end{array}$ 3. $\begin{array}{r} 28 \\ +\ 60 \\ \hline \end{array}$ 4. $\begin{array}{r} 32 \\ +\ 20 \\ \hline \end{array}$ 5. $\begin{array}{r} 67 \\ +\ 21 \\ \hline \end{array}$

6. $\begin{array}{r} 24 \\ +\ 71 \\ \hline \end{array}$ 7. $\begin{array}{r} 35 \\ +\ 14 \\ \hline \end{array}$ 8. $\begin{array}{r} 56 \\ +\ 34 \\ \hline \end{array}$ 9. $\begin{array}{r} 28 \\ +\ 51 \\ \hline \end{array}$ 10. $\begin{array}{r} 50 \\ +\ 23 \\ \hline \end{array}$

11. $\begin{array}{r} 59 \\ +\ 31 \\ \hline \end{array}$ 12. $\begin{array}{r} 27 \\ +\ 23 \\ \hline \end{array}$ 13. $\begin{array}{r} 75 \\ +\ 17 \\ \hline \end{array}$ 14. $\begin{array}{r} 67 \\ +\ 17 \\ \hline \end{array}$ 15. $\begin{array}{r} 38 \\ +\ 52 \\ \hline \end{array}$

16. $\begin{array}{r} 38 \\ +\ 27 \\ \hline \end{array}$ 17. $\begin{array}{r} 45 \\ +\ 48 \\ \hline \end{array}$ 18. $\begin{array}{r} 58 \\ +\ 28 \\ \hline \end{array}$ 19. $\begin{array}{r} 35 \\ +\ 45 \\ \hline \end{array}$ 20. $\begin{array}{r} 35 \\ +\ 16 \\ \hline \end{array}$

21. $\begin{array}{r} 44 \\ +\ 74 \\ \hline \end{array}$ 22. $\begin{array}{r} 72 \\ +\ 32 \\ \hline \end{array}$ 23. $\begin{array}{r} 43 \\ +\ 65 \\ \hline \end{array}$ 24. $\begin{array}{r} 74 \\ +\ 71 \\ \hline \end{array}$ 25. $\begin{array}{r} 33 \\ +\ 76 \\ \hline \end{array}$

26. $\begin{array}{r} 72 \\ +\ 83 \\ \hline \end{array}$ 27. $\begin{array}{r} 63 \\ +\ 52 \\ \hline \end{array}$ 28. $\begin{array}{r} 85 \\ +\ 44 \\ \hline \end{array}$ 29. $\begin{array}{r} 23 \\ +\ 84 \\ \hline \end{array}$ 30. $\begin{array}{r} 96 \\ +\ 80 \\ \hline \end{array}$

Name: **Lesson 9-4 Adding two digit numbers**

---

1. 53 + 41
2. 84 + 11
3. 33 + 57
4. 94 + 15
5. 65 + 23
6. 74 + 16
7. 43 + 55
8. 54 + 24
9. 36 + 51
10. 24 + 42
11. 67 + 23
12. 37 + 25
13. 23 + 39
14. 77 + 14
15. 69 + 17
16. 89 + 22
17. 69 + 31
18. 46 + 46
19. 59 + 21
20. 49 + 18
21. 93 + 42
22. 92 + 31
23. 93 + 73
24. 27 + 89
25. 45 + 62
26. 79 + 97
27. 78 + 62
28. 95 + 33
29. 63 + 56
30. 73 + 67

Name: **Lesson 9-5 Adding two digit numbers**

1. $63 + 26$	2. $56 + 13$	3. $82 + 23$	4. $82 + 32$	5. $47 + 33$
6. $25 + 44$	7. $32 + 27$	8. $53 + 31$	9. $53 + 53$	10. $54 + 21$
11. $36 + 34$	12. $33 + 39$	13. $37 + 43$	14. $37 + 43$	15. $67 + 27$
16. $86 + 18$	17. $67 + 16$	18. $45 + 36$	19. $45 + 36$	20. $69 + 29$
21. $85 + 64$	22. $32 + 75$	23. $85 + 41$	24. $85 + 41$	25. $94 + 13$
26. $69 + 71$	27. $32 + 82$	28. $76 + 81$	29. $76 + 92$	30. $43 + 92$

Name: **Lesson 9-6 Adding two digit numbers**

1. $47 + 31$	2. $66 + 24$	3. $35 + 34$	4. $35 + 64$	5. $16 + 61$
6. $43 + 22$	7. $48 + 12$	8. $43 + 15$	9. $93 + 23$	10. $45 + 75$
11. $73 + 39$	12. $46 + 15$	13. $45 + 56$	14. $69 + 17$	15. $34 + 67$
16. $33 + 69$	17. $79 + 28$	18. $68 + 35$	19. $45 + 29$	20. $36 + 24$
21. $72 + 34$	22. $84 + 72$	23. $63 + 92$	24. $28 + 92$	25. $82 + 33$
26. $43 + 72$	27. $87 + 83$	28. $75 + 63$	29. $93 + 95$	30. $45 + 93$

Name: Lesson 9-7 Adding two digit numbers

1. 75 + 13

2. 75 + 44

3. 23 + 74

4. 54 + 24

5. 41 + 93

6. 26 + 31

7. 84 + 45

8. 26 + 44

9. 23 + 49

10. 32 + 48

11. 35 + 15

12. 30 + 29

13. 39 + 29

14. 45 + 65

15. 69 + 96

16. 38 + 16

17. 59 + 24

18. 79 + 34

19. 69 + 23

20. 28 + 75

21. 24 + 93

22. 94 + 47

23. 33 + 97

24. 99 + 99

25. 64 + 44

26. 83 + 85

27. 69 + 62

28. 68 + 43

29. 78 + 87

30. 48 + 49

Name: Lesson 9-8 Adding two digit numbers

1. $\begin{array}{r} 34 \\ +\ 24 \\ \hline \end{array}$

2. $\begin{array}{r} 25 \\ +\ 15 \\ \hline \end{array}$

3. $\begin{array}{r} 46 \\ +\ 14 \\ \hline \end{array}$

4. $\begin{array}{r} 33 \\ +\ 77 \\ \hline \end{array}$

5. $\begin{array}{r} 54 \\ +\ 25 \\ \hline \end{array}$

6. $\begin{array}{r} 43 \\ +\ 67 \\ \hline \end{array}$

7. $\begin{array}{r} 84 \\ +\ 23 \\ \hline \end{array}$

8. $\begin{array}{r} 62 \\ +\ 64 \\ \hline \end{array}$

9. $\begin{array}{r} 27 \\ +\ 36 \\ \hline \end{array}$

10. $\begin{array}{r} 39 \\ +\ 42 \\ \hline \end{array}$

11. $\begin{array}{r} 46 \\ +\ 24 \\ \hline \end{array}$

12. $\begin{array}{r} 58 \\ +\ 32 \\ \hline \end{array}$

13. $\begin{array}{r} 28 \\ +\ 76 \\ \hline \end{array}$

14. $\begin{array}{r} 63 \\ +\ 29 \\ \hline \end{array}$

15. $\begin{array}{r} 65 \\ +\ 25 \\ \hline \end{array}$

16. $\begin{array}{r} 47 \\ +\ 26 \\ \hline \end{array}$

17. $\begin{array}{r} 79 \\ +\ 38 \\ \hline \end{array}$

18. $\begin{array}{r} 34 \\ +\ 38 \\ \hline \end{array}$

19. $\begin{array}{r} 28 \\ +\ 65 \\ \hline \end{array}$

20. $\begin{array}{r} 83 \\ +\ 27 \\ \hline \end{array}$

21. $\begin{array}{r} 63 \\ +\ 97 \\ \hline \end{array}$

22. $\begin{array}{r} 53 \\ +\ 82 \\ \hline \end{array}$

23. $\begin{array}{r} 57 \\ +\ 52 \\ \hline \end{array}$

24. $\begin{array}{r} 45 \\ +\ 84 \\ \hline \end{array}$

25. $\begin{array}{r} 91 \\ +\ 19 \\ \hline \end{array}$

26. $\begin{array}{r} 93 \\ +\ 72 \\ \hline \end{array}$

27. $\begin{array}{r} 99 \\ +\ 32 \\ \hline \end{array}$

28. $\begin{array}{r} 47 \\ +\ 73 \\ \hline \end{array}$

29. $\begin{array}{r} 92 \\ +\ 82 \\ \hline \end{array}$

30. $\begin{array}{r} 44 \\ +\ 76 \\ \hline \end{array}$

Name: **Lesson 9-9 Adding two digit numbers**

1. 34 + 36

2. 62 + 36

3. 23 + 44

4. 73 + 37

5. 41 + 72

6. 56 + 42

7. 14 + 76

8. 55 + 46

9. 82 + 29

10. 22 + 83

11. 78 + 12

12. 23 + 99

13. 29 + 38

14. 45 + 27

15. 25 + 75

16. 47 + 33

17. 36 + 68

18. 47 + 37

19. 48 + 84

20. 43 + 39

21. 59 + 51

22. 82 + 62

23. 43 + 98

24. 36 + 77

25. 91 + 93

26. 33 + 77

27. 63 + 77

28. 96 + 95

29. 35 + 75

30. 36 + 97

Name: **Lesson 9-10 Adding two digit numbers**

1. $\begin{array}{r} 24 \\ +\ 46 \\ \hline \end{array}$
2. $\begin{array}{r} 76 \\ +\ 13 \\ \hline \end{array}$
3. $\begin{array}{r} 47 \\ +\ 27 \\ \hline \end{array}$
4. $\begin{array}{r} 54 \\ +\ 54 \\ \hline \end{array}$
5. $\begin{array}{r} 84 \\ +\ 25 \\ \hline \end{array}$
6. $\begin{array}{r} 35 \\ +\ 41 \\ \hline \end{array}$
7. $\begin{array}{r} 67 \\ +\ 26 \\ \hline \end{array}$
8. $\begin{array}{r} 25 \\ +\ 26 \\ \hline \end{array}$
9. $\begin{array}{r} 62 \\ +\ 23 \\ \hline \end{array}$
10. $\begin{array}{r} 41 \\ +\ 92 \\ \hline \end{array}$
11. $\begin{array}{r} 57 \\ +\ 75 \\ \hline \end{array}$
12. $\begin{array}{r} 39 \\ +\ 31 \\ \hline \end{array}$
13. $\begin{array}{r} 49 \\ +\ 28 \\ \hline \end{array}$
14. $\begin{array}{r} 36 \\ +\ 24 \\ \hline \end{array}$
15. $\begin{array}{r} 59 \\ +\ 33 \\ \hline \end{array}$
16. $\begin{array}{r} 49 \\ +\ 49 \\ \hline \end{array}$
17. $\begin{array}{r} 33 \\ +\ 78 \\ \hline \end{array}$
18. $\begin{array}{r} 38 \\ +\ 55 \\ \hline \end{array}$
19. $\begin{array}{r} 52 \\ +\ 58 \\ \hline \end{array}$
20. $\begin{array}{r} 78 \\ +\ 17 \\ \hline \end{array}$
21. $\begin{array}{r} 25 \\ +\ 98 \\ \hline \end{array}$
22. $\begin{array}{r} 42 \\ +\ 29 \\ \hline \end{array}$
23. $\begin{array}{r} 23 \\ +\ 97 \\ \hline \end{array}$
24. $\begin{array}{r} 63 \\ +\ 69 \\ \hline \end{array}$
25. $\begin{array}{r} 53 \\ +\ 46 \\ \hline \end{array}$
26. $\begin{array}{r} 78 \\ +\ 87 \\ \hline \end{array}$
27. $\begin{array}{r} 75 \\ +\ 77 \\ \hline \end{array}$
28. $\begin{array}{r} 73 \\ +\ 38 \\ \hline \end{array}$
29. $\begin{array}{r} 95 \\ +\ 96 \\ \hline \end{array}$
30. $\begin{array}{r} 85 \\ +\ 56 \\ \hline \end{array}$

Name: **Lesson 10-1 Subtracting two digit numbers**

---

1. $\begin{array}{r} 72 \\ -\ 14 \\ \hline \end{array}$	2. $\begin{array}{r} 53 \\ -\ 27 \\ \hline \end{array}$	3. $\begin{array}{r} 84 \\ -\ 36 \\ \hline \end{array}$	4. $\begin{array}{r} 96 \\ -\ 58 \\ \hline \end{array}$	5. $\begin{array}{r} 44 \\ -\ 19 \\ \hline \end{array}$
6. $\begin{array}{r} 96 \\ -\ 68 \\ \hline \end{array}$	7. $\begin{array}{r} 62 \\ -\ 24 \\ \hline \end{array}$	8. $\begin{array}{r} 98 \\ -\ 39 \\ \hline \end{array}$	9. $\begin{array}{r} 72 \\ -\ 39 \\ \hline \end{array}$	10. $\begin{array}{r} 59 \\ -\ 26 \\ \hline \end{array}$
11. $\begin{array}{r} 91 \\ -\ 13 \\ \hline \end{array}$	12. $\begin{array}{r} 85 \\ -\ 31 \\ \hline \end{array}$	13. $\begin{array}{r} 56 \\ -\ 28 \\ \hline \end{array}$	14. $\begin{array}{r} 97 \\ -\ 58 \\ \hline \end{array}$	15. $\begin{array}{r} 73 \\ -\ 52 \\ \hline \end{array}$
16. $\begin{array}{r} 85 \\ -\ 36 \\ \hline \end{array}$	17. $\begin{array}{r} 81 \\ -\ 68 \\ \hline \end{array}$	18. $\begin{array}{r} 53 \\ -\ 16 \\ \hline \end{array}$	19. $\begin{array}{r} 74 \\ -\ 25 \\ \hline \end{array}$	20. $\begin{array}{r} 89 \\ -\ 38 \\ \hline \end{array}$
21. $\begin{array}{r} 82 \\ -\ 37 \\ \hline \end{array}$	22. $\begin{array}{r} 75 \\ -\ 44 \\ \hline \end{array}$	23. $\begin{array}{r} 62 \\ -\ 34 \\ \hline \end{array}$	24. $\begin{array}{r} 81 \\ -\ 66 \\ \hline \end{array}$	25. $\begin{array}{r} 53 \\ -\ 35 \\ \hline \end{array}$
26. $\begin{array}{r} 31 \\ -\ 17 \\ \hline \end{array}$	27. $\begin{array}{r} 84 \\ -\ 56 \\ \hline \end{array}$	28. $\begin{array}{r} 74 \\ -\ 49 \\ \hline \end{array}$	29. $\begin{array}{r} 68 \\ -\ 29 \\ \hline \end{array}$	30. $\begin{array}{r} 62 \\ -\ 36 \\ \hline \end{array}$

Name: Lesson 10-2 Subtracting two digit numbers

1. $95 - 46$	2. $82 - 32$	3. $93 - 13$	4. $52 - 25$	5. $61 - 17$
6. $88 - 59$	7. $83 - 15$	8. $98 - 79$	9. $86 - 19$	10. $84 - 49$
11. $93 - 17$	12. $94 - 47$	13. $77 - 29$	14. $99 - 24$	15. $55 - 37$
16. $45 - 28$	17. $75 - 28$	18. $47 - 18$	19. $42 - 17$	20. $91 - 48$
21. $73 - 55$	22. $65 - 49$	23. $76 - 38$	24. $53 - 27$	25. $45 - 17$
26. $95 - 56$	27. $72 - 58$	28. $52 - 16$	29. $65 - 38$	30. $81 - 15$

Name:

Lesson 10-3 Subtracting two digit numbers

---

1. $82 - 17$	2. $71 - 33$	3. $62 - 24$	4. $54 - 28$	5. $83 - 45$
6. $83 - 65$	7. $77 - 49$	8. $82 - 35$	9. $73 - 27$	10. $44 - 17$
11. $57 - 18$	12. $73 - 36$	13. $84 - 47$	14. $71 - 18$	15. $82 - 23$
16. $51 - 19$	17. $95 - 58$	18. $31 - 15$	19. $43 - 26$	20. $64 - 39$
21. $71 - 27$	22. $63 - 34$	23. $65 - 27$	24. $66 - 38$	25. $42 - 18$
26. $54 - 16$	27. $46 - 17$	28. $87 - 49$	29. $52 - 37$	30. $92 - 57$

Name: **Lesson 10-4 Subtracting two digit numbers**

1. $\begin{array}{r} 71 \\ -\ 58 \\ \hline \end{array}$	2. $\begin{array}{r} 83 \\ -\ 36 \\ \hline \end{array}$	3. $\begin{array}{r} 81 \\ -\ 44 \\ \hline \end{array}$	4. $\begin{array}{r} 93 \\ -\ 57 \\ \hline \end{array}$	5. $\begin{array}{r} 61 \\ -\ 22 \\ \hline \end{array}$
6. $\begin{array}{r} 53 \\ -\ 19 \\ \hline \end{array}$	7. $\begin{array}{r} 95 \\ -\ 47 \\ \hline \end{array}$	8. $\begin{array}{r} 52 \\ -\ 15 \\ \hline \end{array}$	9. $\begin{array}{r} 97 \\ -\ 65 \\ \hline \end{array}$	10. $\begin{array}{r} 55 \\ -\ 38 \\ \hline \end{array}$
11. $\begin{array}{r} 64 \\ -\ 37 \\ \hline \end{array}$	12. $\begin{array}{r} 88 \\ -\ 59 \\ \hline \end{array}$	13. $\begin{array}{r} 85 \\ -\ 49 \\ \hline \end{array}$	14. $\begin{array}{r} 47 \\ -\ 19 \\ \hline \end{array}$	15. $\begin{array}{r} 31 \\ -\ 13 \\ \hline \end{array}$
16. $\begin{array}{r} 74 \\ -\ 48 \\ \hline \end{array}$	17. $\begin{array}{r} 62 \\ -\ 34 \\ \hline \end{array}$	18. $\begin{array}{r} 97 \\ -\ 79 \\ \hline \end{array}$	19. $\begin{array}{r} 65 \\ -\ 15 \\ \hline \end{array}$	20. $\begin{array}{r} 48 \\ -\ 14 \\ \hline \end{array}$
21. $\begin{array}{r} 72 \\ -\ 59 \\ \hline \end{array}$	22. $\begin{array}{r} 61 \\ -\ 23 \\ \hline \end{array}$	23. $\begin{array}{r} 77 \\ -\ 48 \\ \hline \end{array}$	24. $\begin{array}{r} 82 \\ -\ 51 \\ \hline \end{array}$	25. $\begin{array}{r} 84 \\ -\ 38 \\ \hline \end{array}$
26. $\begin{array}{r} 85 \\ -\ 38 \\ \hline \end{array}$	27. $\begin{array}{r} 72 \\ -\ 39 \\ \hline \end{array}$	28. $\begin{array}{r} 73 \\ -\ 16 \\ \hline \end{array}$	29. $\begin{array}{r} 82 \\ -\ 44 \\ \hline \end{array}$	30. $\begin{array}{r} 57 \\ -\ 28 \\ \hline \end{array}$

Name: **Lesson 10-5 Subtracting two digit numbers**

1. $\begin{array}{r} 62 \\ -\ 23 \\ \hline \end{array}$	2. $\begin{array}{r} 81 \\ -\ 46 \\ \hline \end{array}$	3. $\begin{array}{r} 55 \\ -\ 29 \\ \hline \end{array}$	4. $\begin{array}{r} 45 \\ -\ 17 \\ \hline \end{array}$	5. $\begin{array}{r} 83 \\ -\ 29 \\ \hline \end{array}$
6. $\begin{array}{r} 54 \\ -\ 19 \\ \hline \end{array}$	7. $\begin{array}{r} 83 \\ -\ 24 \\ \hline \end{array}$	8. $\begin{array}{r} 67 \\ -\ 48 \\ \hline \end{array}$	9. $\begin{array}{r} 33 \\ -\ 16 \\ \hline \end{array}$	10. $\begin{array}{r} 85 \\ -\ 58 \\ \hline \end{array}$
11. $\begin{array}{r} 46 \\ -\ 14 \\ \hline \end{array}$	12. $\begin{array}{r} 73 \\ -\ 29 \\ \hline \end{array}$	13. $\begin{array}{r} 65 \\ -\ 26 \\ \hline \end{array}$	14. $\begin{array}{r} 52 \\ -\ 26 \\ \hline \end{array}$	15. $\begin{array}{r} 75 \\ -\ 37 \\ \hline \end{array}$
16. $\begin{array}{r} 53 \\ -\ 25 \\ \hline \end{array}$	17. $\begin{array}{r} 71 \\ -\ 38 \\ \hline \end{array}$	18. $\begin{array}{r} 73 \\ -\ 46 \\ \hline \end{array}$	19. $\begin{array}{r} 86 \\ -\ 27 \\ \hline \end{array}$	20. $\begin{array}{r} 88 \\ -\ 49 \\ \hline \end{array}$
21. $\begin{array}{r} 73 \\ -\ 36 \\ \hline \end{array}$	22. $\begin{array}{r} 55 \\ -\ 26 \\ \hline \end{array}$	23. $\begin{array}{r} 82 \\ -\ 48 \\ \hline \end{array}$	24. $\begin{array}{r} 71 \\ -\ 34 \\ \hline \end{array}$	25. $\begin{array}{r} 63 \\ -\ 28 \\ \hline \end{array}$
26. $\begin{array}{r} 95 \\ -\ 57 \\ \hline \end{array}$	27. $\begin{array}{r} 56 \\ -\ 17 \\ \hline \end{array}$	28. $\begin{array}{r} 64 \\ -\ 19 \\ \hline \end{array}$	29. $\begin{array}{r} 61 \\ -\ 25 \\ \hline \end{array}$	30. $\begin{array}{r} 62 \\ -\ 34 \\ \hline \end{array}$

Name: **Lesson 10-6 Subtracting two digit numbers**

---

1. $\begin{array}{r} 85 \\ -\ 56 \\ \hline \end{array}$	2. $\begin{array}{r} 66 \\ -\ 29 \\ \hline \end{array}$	3. $\begin{array}{r} 92 \\ -\ 55 \\ \hline \end{array}$	4. $\begin{array}{r} 81 \\ -\ 14 \\ \hline \end{array}$	5. $\begin{array}{r} 55 \\ -\ 19 \\ \hline \end{array}$
6. $\begin{array}{r} 62 \\ -\ 16 \\ \hline \end{array}$	7. $\begin{array}{r} 92 \\ -\ 37 \\ \hline \end{array}$	8. $\begin{array}{r} 84 \\ -\ 48 \\ \hline \end{array}$	9. $\begin{array}{r} 75 \\ -\ 28 \\ \hline \end{array}$	10. $\begin{array}{r} 94 \\ -\ 57 \\ \hline \end{array}$
11. $\begin{array}{r} 91 \\ -\ 42 \\ \hline \end{array}$	12. $\begin{array}{r} 64 \\ -\ 49 \\ \hline \end{array}$	13. $\begin{array}{r} 33 \\ -\ 16 \\ \hline \end{array}$	14. $\begin{array}{r} 96 \\ -\ 68 \\ \hline \end{array}$	15. $\begin{array}{r} 42 \\ -\ 16 \\ \hline \end{array}$
16. $\begin{array}{r} 66 \\ -\ 39 \\ \hline \end{array}$	17. $\begin{array}{r} 82 \\ -\ 55 \\ \hline \end{array}$	18. $\begin{array}{r} 45 \\ -\ 26 \\ \hline \end{array}$	19. $\begin{array}{r} 81 \\ -\ 47 \\ \hline \end{array}$	20. $\begin{array}{r} 73 \\ -\ 19 \\ \hline \end{array}$
21. $\begin{array}{r} 84 \\ -\ 65 \\ \hline \end{array}$	22. $\begin{array}{r} 54 \\ -\ 29 \\ \hline \end{array}$	23. $\begin{array}{r} 72 \\ -\ 38 \\ \hline \end{array}$	24. $\begin{array}{r} 45 \\ -\ 19 \\ \hline \end{array}$	25. $\begin{array}{r} 97 \\ -\ 69 \\ \hline \end{array}$
26. $\begin{array}{r} 86 \\ -\ 18 \\ \hline \end{array}$	27. $\begin{array}{r} 93 \\ -\ 79 \\ \hline \end{array}$	28. $\begin{array}{r} 92 \\ -\ 54 \\ \hline \end{array}$	29. $\begin{array}{r} 64 \\ -\ 27 \\ \hline \end{array}$	30. $\begin{array}{r} 76 \\ -\ 48 \\ \hline \end{array}$

Name:

Lesson 10-7 Subtracting two digit numbers

---

1. $\begin{array}{r} 76 \\ -\ 59 \\ \hline \end{array}$
2. $\begin{array}{r} 52 \\ -\ 18 \\ \hline \end{array}$
3. $\begin{array}{r} 51 \\ -\ 26 \\ \hline \end{array}$
4. $\begin{array}{r} 68 \\ -\ 19 \\ \hline \end{array}$
5. $\begin{array}{r} 84 \\ -\ 37 \\ \hline \end{array}$
6. $\begin{array}{r} 73 \\ -\ 37 \\ \hline \end{array}$
7. $\begin{array}{r} 95 \\ -\ 19 \\ \hline \end{array}$
8. $\begin{array}{r} 97 \\ -\ 39 \\ \hline \end{array}$
9. $\begin{array}{r} 42 \\ -\ 16 \\ \hline \end{array}$
10. $\begin{array}{r} 66 \\ -\ 28 \\ \hline \end{array}$
11. $\begin{array}{r} 84 \\ -\ 48 \\ \hline \end{array}$
12. $\begin{array}{r} 77 \\ -\ 39 \\ \hline \end{array}$
13. $\begin{array}{r} 63 \\ -\ 36 \\ \hline \end{array}$
14. $\begin{array}{r} 54 \\ -\ 29 \\ \hline \end{array}$
15. $\begin{array}{r} 63 \\ -\ 19 \\ \hline \end{array}$
16. $\begin{array}{r} 94 \\ -\ 17 \\ \hline \end{array}$
17. $\begin{array}{r} 63 \\ -\ 25 \\ \hline \end{array}$
18. $\begin{array}{r} 33 \\ -\ 18 \\ \hline \end{array}$
19. $\begin{array}{r} 91 \\ -\ 52 \\ \hline \end{array}$
20. $\begin{array}{r} 55 \\ -\ 28 \\ \hline \end{array}$
21. $\begin{array}{r} 73 \\ -\ 49 \\ \hline \end{array}$
22. $\begin{array}{r} 61 \\ -\ 33 \\ \hline \end{array}$
23. $\begin{array}{r} 94 \\ -\ 39 \\ \hline \end{array}$
24. $\begin{array}{r} 97 \\ -\ 68 \\ \hline \end{array}$
25. $\begin{array}{r} 91 \\ -\ 36 \\ \hline \end{array}$
26. $\begin{array}{r} 45 \\ -\ 18 \\ \hline \end{array}$
27. $\begin{array}{r} 92 \\ -\ 35 \\ \hline \end{array}$
28. $\begin{array}{r} 86 \\ -\ 48 \\ \hline \end{array}$
29. $\begin{array}{r} 95 \\ -\ 66 \\ \hline \end{array}$
30. $\begin{array}{r} 86 \\ -\ 29 \\ \hline \end{array}$

Name: **Lesson 10-8 Subtracting two digit numbers**

1. 85 − 66	2. 53 − 18	3. 81 − 55	4. 62 − 27	5. 92 − 39
6. 73 − 15	7. 89 − 49	8. 93 − 16	9. 75 − 39	10. 72 − 44
11. 82 − 28	12. 75 − 68	13. 61 − 15	14. 94 − 36	15. 82 − 33
16. 95 − 39	17. 93 − 68	18. 72 − 16	19. 81 − 14	20. 94 − 69
21. 71 − 46	22. 92 − 25	23. 55 − 18	24. 86 − 59	25. 62 − 17
26. 55 − 27	27. 87 − 58	28. 67 − 29	29. 92 − 28	30. 93 − 69

Name: **Lesson 10-9 Subtracting two digit numbers**

1. $72 - 25$	2. $85 - 48$	3. $93 - 29$	4. $93 - 41$	5. $43 - 26$
6. $55 - 39$	7. $63 - 15$	8. $97 - 39$	9. $92 - 57$	10. $72 - 33$
11. $65 - 28$	12. $53 - 19$	13. $84 - 55$	14. $81 - 58$	15. $62 - 44$
16. $92 - 44$	17. $98 - 38$	18. $85 - 28$	19. $56 - 17$	20. $94 - 59$
21. $95 - 16$	22. $95 - 27$	23. $73 - 49$	24. $71 - 34$	25. $83 - 39$
26. $83 - 17$	27. $85 - 39$	28. $64 - 27$	29. $91 - 46$	30. $93 - 69$

Name: **Lesson 10-10 Subtracting two digit numbers**

1. $65 - 26$	2. $72 - 27$	3. $92 - 28$	4. $83 - 27$	5. $95 - 68$
6. $73 - 35$	7. $75 - 49$	8. $95 - 39$	9. $65 - 36$	10. $58 - 19$
11. $93 - 28$	12. $63 - 25$	13. $64 - 25$	14. $57 - 19$	15. $74 - 58$
16. $92 - 59$	17. $55 - 29$	18. $65 - 39$	19. $86 - 37$	20. $91 - 66$
21. $65 - 27$	22. $51 - 28$	23. $51 - 27$	24. $52 - 14$	25. $93 - 48$
26. $46 - 19$	27. $62 - 47$	28. $82 - 46$	29. $71 - 33$	30. $55 - 27$

1. $12 + 15 + 1 =$
2. $14 + 14 + 2 =$
3. $12 + 19 + 3 =$
4. $11 + 12 + 1 =$
5. $12 + 16 + 2 =$
6. $18 + 14 + 3 =$
7. $13 + 14 + 1 =$
8. $16 + 16 + 2 =$
9. $13 + 12 + 3 =$
10. $17 + 11 + 1 =$
11. $13 + 14 + 2 =$
12. $15 + 15 + 3 =$
13. $15 + 13 + 1 =$
14. $13 + 17 + 2 =$
15. $14 + 17 + 3 =$
16. $11 + 16 + 1 =$
17. $15 + 17 + 2 =$
18. $19 + 19 + 3 =$
19. $14 + 17 + 1 =$
20. $14 + 18 + 2 =$
21. $14 + 15 + 3 =$
22. $18 + 16 + 1 =$
23. $19 + 17 + 2 =$
24. $12 + 14 + 3 =$
25. $12 + 19 + 1 =$
26. $18 + 13 + 2 =$
27. $12 + 19 + 3 =$
28. $13 + 17 + 1 =$
29. $16 + 15 + 2 =$
30. $15 + 17 + 3 =$

Name: **Lesson 11-2** **Adding three numbers**

1. $11 + 15 + 2 =$
2. $17 + 13 + 3 =$
3. $12 + 18 + 4 =$
4. $13 + 15 + 2 =$
5. $18 + 19 + 3 =$
6. $13 + 17 + 4 =$
7. $18 + 17 + 2 =$
8. $13 + 15 + 3 =$
9. $14 + 17 + 4 =$
10. $14 + 15 + 2 =$
11. $18 + 19 + 3 =$
12. $16 + 17 + 4 =$
13. $15 + 19 + 2 =$
14. $12 + 17 + 3 =$
15. $13 + 16 + 4 =$
16. $16 + 14 + 2 =$
17. $18 + 13 + 3 =$
18. $15 + 19 + 4 =$
19. $13 + 17 + 2 =$
20. $12 + 16 + 3 =$
21. $16 + 18 + 4 =$
22. $12 + 18 + 2 =$
23. $15 + 15 + 3 =$
24. $11 + 19 + 4 =$
25. $17 + 17 + 2 =$
26. $19 + 15 + 3 =$
27. $14 + 16 + 4 =$
28. $16 + 15 + 2 =$
29. $16 + 16 + 3 =$
30. $19 + 19 + 4 =$

Name: **Lesson 11-3** **Adding three numbers**

1. $16 + 11 + 3 =$	2. $14 + 14 + 4 =$	3. $11 + 19 + 5 =$
4. $12 + 14 + 3 =$	5. $17 + 11 + 4 =$	6. $13 + 14 + 5 =$
7. $11 + 13 + 3 =$	8. $12 + 15 + 4 =$	9. $11 + 12 + 5 =$
10. $13 + 12 + 3 =$	11. $14 + 13 + 4 =$	12. $14 + 11 + 5 =$
13. $11 + 15 + 3 =$	14. $17 + 12 + 4 =$	15. $15 + 14 + 5 =$
16. $13 + 16 + 3 =$	17. $14 + 12 + 4 =$	18. $12 + 13 + 5 =$
19. $17 + 11 + 3 =$	20. $16 + 12 + 4 =$	21. $12 + 17 + 5 =$
22. $13 + 15 + 3 =$	23. $11 + 14 + 4 =$	24. $11 + 16 + 5 =$
25. $12 + 11 + 3 =$	26. $15 + 13 + 4 =$	27. $13 + 13 + 5 =$
28. $15 + 19 + 3 =$	29. $18 + 11 + 4 =$	30. $14 + 15 + 5 =$

1. $13 + 13 + 4 =$	2. $16 + 11 + 5 =$	3. $12 + 14 + 6 =$
4. $12 + 11 + 4 =$	5. $11 + 15 + 5 =$	6. $14 + 11 + 6 =$
7. $18 + 11 + 4 =$	8. $12 + 13 + 5 =$	9. $11 + 18 + 6 =$
10. $16 + 13 + 4 =$	11. $14 + 15 + 5 =$	12. $13 + 14 + 6 =$
13. $14 + 13 + 4 =$	14. $12 + 16 + 5 =$	15. $15 + 16 + 6 =$
16. $11 + 17 + 4 =$	17. $17 + 12 + 5 =$	18. $15 + 19 + 6 =$
19. $11 + 14 + 4 =$	20. $13 + 14 + 5 =$	21. $14 + 14 + 6 =$
22. $16 + 12 + 4 =$	23. $12 + 18 + 5 =$	24. $13 + 16 + 6 =$
25. $15 + 14 + 4 =$	26. $11 + 19 + 5 =$	27. $17 + 18 + 6 =$
28. $13 + 15 + 4 =$	29. $15 + 16 + 5 =$	30. $14 + 19 + 6 =$

1. $18 + 11 + 5 =$	2. $15 + 12 + 6 =$	3. $14 + 11 + 7 =$
4. $14 + 12 + 5 =$	5. $13 + 16 + 6 =$	6. $12 + 17 + 7 =$
7. $11 + 15 + 5 =$	8. $11 + 14 + 6 =$	9. $11 + 17 + 7 =$
10. $14 + 13 + 5 =$	11. $13 + 11 + 6 =$	12. $12 + 16 + 7 =$
13. $11 + 17 + 5 =$	14. $12 + 13 + 6 =$	15. $15 + 15 + 7 =$
16. $12 + 15 + 5 =$	17. $13 + 15 + 6 =$	18. $16 + 13 + 7 =$
19. $13 + 13 + 5 =$	20. $12 + 18 + 6 =$	21. $14 + 12 + 7 =$
22. $14 + 15 + 5 =$	23. $13 + 18 + 6 =$	24. $15 + 16 + 7 =$
25. $17 + 13 + 5 =$	26. $14 + 14 + 6 =$	27. $17 + 18 + 7 =$
28. $16 + 17 + 5 =$	29. $15 + 17 + 6 =$	30. $16 + 19 + 7 =$

1. $13 + 12 + 6 =$	2. $11 + 18 + 7 =$	3. $13 + 11 + 8 =$
4. $15 + 11 + 6 =$	5. $12 + 12 + 7 =$	6. $11 + 16 + 8 =$
7. $14 + 13 + 6 =$	8. $16 + 12 + 7 =$	9. $13 + 13 + 8 =$
10. $12 + 15 + 6 =$	11. $17 + 11 + 7 =$	12. $14 + 11 + 8 =$
13. $15 + 14 + 6 =$	14. $15 + 12 + 7 =$	15. $13 + 16 + 8 =$
16. $14 + 14 + 6 =$	17. $12 + 14 + 7 =$	18. $13 + 17 + 8 =$
19. $17 + 12 + 6 =$	20. $15 + 13 + 7 =$	21. $18 + 11 + 8 =$
22. $16 + 13 + 6 =$	23. $11 + 15 + 7 =$	24. $12 + 17 + 8 =$
25. $11 + 14 + 6 =$	26. $12 + 16 + 7 =$	27. $14 + 15 + 8 =$
28. $17 + 15 + 6 =$	29. $13 + 14 + 7 =$	30. $14 + 19 + 8 =$

1. $18 + 11 + 7 =$	2. $14 + 13 + 8 =$	3. $11 + 12 + 9 =$
4. $11 + 15 + 7 =$	5. $17 + 11 + 8 =$	6. $13 + 12 + 9 =$
7. $12 + 14 + 7 =$	8. $11 + 12 + 8 =$	9. $15 + 12 + 9 =$
10. $16 + 13 + 7 =$	11. $16 + 11 + 8 =$	12. $14 + 15 + 9 =$
13. $15 + 12 + 7 =$	14. $11 + 14 + 8 =$	15. $17 + 12 + 9 =$
16. $14 + 13 + 7 =$	17. $12 + 17 + 8 =$	18. $13 + 13 + 9 =$
19. $12 + 19 + 7 =$	20. $14 + 14 + 8 =$	21. $12 + 16 + 9 =$
22. $14 + 11 + 7 =$	23. $12 + 19 + 8 =$	24. $11 + 17 + 9 =$
25. $15 + 14 + 7 =$	26. $13 + 15 + 8 =$	27. $19 + 12 + 9 =$
28. $18 + 17 + 7 =$	29. $12 + 18 + 8 =$	30. $13 + 17 + 9 =$

Name: Lesson 11-8 Adding three numbers

1. $14 + 11 + 8 =$	2. $16 + 13 + 9 =$	3. $11 + 15 + 10 =$
4. $15 + 14 + 8 =$	5. $12 + 14 + 9 =$	6. $13 + 14 + 10 =$
7. $16 + 11 + 8 =$	8. $15 + 12 + 9 =$	9. $14 + 14 + 10 =$
10. $13 + 12 + 8 =$	11. $11 + 13 + 9 =$	12. $18 + 11 + 10 =$
13. $17 + 11 + 8 =$	14. $13 + 14 + 9 =$	15. $13 + 16 + 10 =$
16. $12 + 16 + 8 =$	17. $12 + 17 + 9 =$	18. $11 + 14 + 10 =$
19. $15 + 13 + 8 =$	20. $12 + 13 + 9 =$	21. $16 + 12 + 10 =$
22. $14 + 13 + 8 =$	23. $15 + 17 + 9 =$	24. $14 + 15 + 10 =$
25. $12 + 15 + 8 =$	26. $18 + 12 + 9 =$	27. $15 + 11 + 10 =$
28. $15 + 16 + 8 =$	29. $17 + 19 + 9 =$	30. $11 + 17 + 10 =$

Name: **Lesson 11-9 Adding three numbers**

1. $12 + 14 + 1 =$	2. $14 + 15 + 1 =$	3. $12 + 17 + 1 =$
4. $15 + 12 + 2 =$	5. $17 + 11 + 2 =$	6. $13 + 12 + 2 =$
7. $13 + 15 + 3 =$	8. $18 + 11 + 3 =$	9. $12 + 15 + 3 =$
10. $14 + 14 + 4 =$	11. $12 + 16 + 4 =$	12. $11 + 12 + 4 =$
13. $12 + 13 + 5 =$	14. $11 + 14 + 5 =$	15. $11 + 17 + 5 =$
16. $16 + 11 + 6 =$	17. $13 + 13 + 6 =$	18. $13 + 16 + 6 =$
19. $11 + 15 + 7 =$	20. $14 + 13 + 7 =$	21. $15 + 11 + 7 =$
22. $16 + 12 + 8 =$	23. $11 + 18 + 8 =$	24. $11 + 16 + 8 =$
25. $14 + 12 + 9 =$	26. $17 + 12 + 9 =$	27. $14 + 11 + 9 =$
28. $15 + 14 + 10 =$	29. $16 + 18 + 10 =$	30. $13 + 19 + 10 =$

Name: **Lesson 11-10** **Adding three numbers**

1. $13 + 12 + 1 =$	2. $16 + 13 + 1 =$	3. $12 + 12 + 1 =$
4. $15 + 14 + 2 =$	5. $12 + 15 + 2 =$	6. $14 + 13 + 2 =$
7. $17 + 11 + 3 =$	8. $14 + 12 + 3 =$	9. $16 + 11 + 3 =$
10. $15 + 13 + 4 =$	11. $11 + 12 + 4 =$	12. $17 + 13 + 4 =$
13. $11 + 15 + 5 =$	14. $11 + 17 + 5 =$	15. $11 + 18 + 5 =$
16. $14 + 11 + 6 =$	17. $17 + 13 + 6 =$	18. $13 + 16 + 6 =$
19. $18 + 11 + 7 =$	20. $15 + 16 + 7 =$	21. $15 + 15 + 7 =$
22. $13 + 16 + 8 =$	23. $16 + 16 + 8 =$	24. $14 + 19 + 8 =$
25. $13 + 19 + 9 =$	26. $14 + 18 + 9 =$	27. $13 + 18 + 9 =$
28. $17 + 18 + 10 =$	29. $17 + 19 + 10 =$	30. $18 + 19 + 10 =$

Name:

Lesson 12-1 Adding and subtracting three numbers

---

1. $24 - 11 + 1 =$
2. $22 - 16 + 2 =$
3. $28 - 19 + 3 =$
4. $28 - 15 + 1 =$
5. $24 - 15 + 2 =$
6. $26 - 19 + 3 =$
7. $26 - 14 + 1 =$
8. $22 - 13 + 2 =$
9. $23 - 14 + 3 =$
10. $28 - 17 + 1 =$
11. $27 - 19 + 2 =$
12. $26 - 18 + 3 =$
13. $22 - 15 + 1 =$
14. $24 - 19 + 2 =$
15. $21 - 15 + 3 =$
16. $25 - 17 + 1 =$
17. $25 - 18 + 2 =$
18. $16 - 13 + 3 =$
19. $23 - 9 + 1 =$
20. $24 - 17 + 2 =$
21. $27 - 18 + 3 =$
22. $23 - 9 + 1 =$
23. $21 - 18 + 2 =$
24. $23 - 12 + 3 =$
25. $22 - 8 + 1 =$
26. $25 - 19 + 2 =$
27. $21 - 13 + 3 =$
28. $25 - 9 + 1 =$
29. $23 - 17 + 2 =$
30. $25 - 19 + 3 =$

Name: **Lesson 12-2 Adding and subtracting three numbers**

1. $38 - 12 + 2 =$
2. $33 - 16 + 3 =$
3. $33 - 19 + 4 =$
4. $32 - 15 + 2 =$
5. $31 - 14 + 3 =$
6. $36 - 17 + 4 =$
7. $37 - 19 + 2 =$
8. $35 - 16 + 3 =$
9. $38 - 19 + 4 =$
10. $33 - 17 + 2 =$
11. $32 - 13 + 3 =$
12. $34 - 18 + 4 =$
13. $36 - 18 + 2 =$
14. $34 - 17 + 3 =$
15. $34 - 15 + 4 =$
16. $33 - 14 + 2 =$
17. $32 - 14 + 3 =$
18. $36 - 19 + 4 =$
19. $31 - 16 + 2 =$
20. $31 - 18 + 3 =$
21. $31 - 12 + 4 =$
22. $32 - 19 + 2 =$
23. $34 - 16 + 3 =$
24. $31 - 17 + 4 =$
25. $31 - 15 + 2 =$
26. $35 - 18 + 3 =$
27. $35 - 19 + 4 =$
28. $34 - 19 + 2 =$
29. $32 - 16 + 3 =$
30. $37 - 19 + 4 =$

Name: **Lesson 12-3 Adding and subtracting three numbers**

1. $33 - 13 + 3 =$
2. $38 - 19 + 4 =$
3. $34 - 15 + 5 =$
4. $33 - 18 + 3 =$
5. $37 - 19 + 4 =$
6. $31 - 14 + 5 =$
7. $31 - 16 + 3 =$
8. $35 - 17 + 4 =$
9. $33 - 16 + 5 =$
10. $35 - 19 + 3 =$
11. $37 - 18 + 4 =$
12. $33 - 19 + 5 =$
13. $33 - 17 + 3 =$
14. $31 - 15 + 4 =$
15. $36 - 18 + 5 =$
16. $31 - 12 + 3 =$
17. $34 - 16 + 4 =$
18. $32 - 19 + 5 =$
19. $34 - 18 + 3 =$
20. $31 - 13 + 4 =$
21. $31 - 17 + 5 =$
22. $33 - 14 + 3 =$
23. $34 - 17 + 4 =$
24. $36 - 19 + 5 =$
25. $32 - 16 + 3 =$
26. $32 - 18 + 4 =$
27. $35 - 16 + 5 =$
28. $31 - 19 + 3 =$
29. $32 - 13 + 4 =$
30. $35 - 18 + 5 =$

Name: Lesson 12-4 Adding and subtracting three numbers

1. $42 - 12 + 4 =$

2. $44 - 16 + 5 =$

3. $43 - 17 + 6 =$

4. $43 - 14 + 4 =$

5. $43 - 19 + 5 =$

6. $41 - 15 + 6 =$

7. $41 - 12 + 4 =$

8. $44 - 18 + 5 =$

9. $42 - 14 + 6 =$

10. $43 - 15 + 4 =$

11. $48 - 19 + 5 =$

12. $41 - 13 + 6 =$

13. $46 - 17 + 4 =$

14. $41 - 14 + 5 =$

15. $45 - 18 + 6 =$

16. $44 - 17 + 4 =$

17. $42 - 16 + 5 =$

18. $43 - 19 + 6 =$

19. $41 - 18 + 4 =$

20. $43 - 18 + 5 =$

21. $45 - 16 + 6 =$

22. $47 - 19 + 4 =$

23. $43 - 16 + 5 =$

24. $46 - 18 + 6 =$

25. $42 - 13 + 4 =$

26. $41 - 19 + 5 =$

27. $45 - 17 + 6 =$

28. $45 - 19 + 4 =$

29. $42 - 15 + 5 =$

30. $46 - 18 + 6 =$

Name: **Lesson 12-5 Adding and subtracting three numbers**

1. $46 - 16 + 5 =$
2. $44 - 15 + 6 =$
3. $44 - 15 + 7 =$
4. $45 - 17 + 5 =$
5. $47 - 19 + 6 =$
6. $47 - 19 + 7 =$
7. $41 - 13 + 5 =$
8. $42 - 17 + 6 =$
9. $42 - 17 + 7 =$
10. $45 - 18 + 5 =$
11. $44 - 19 + 6 =$
12. $44 - 19 + 7 =$
13. $44 - 17 + 5 =$
14. $41 - 16 + 6 =$
15. $41 - 16 + 7 =$
16. $43 - 16 + 5 =$
17. $43 - 19 + 6 =$
18. $43 - 19 + 7 =$
19. $42 - 15 + 5 =$
20. $44 - 16 + 6 =$
21. $44 - 16 + 7 =$
22. $42 - 19 + 5 =$
23. $42 - 13 + 6 =$
24. $42 - 13 + 7 =$
25. $41 - 18 + 5 =$
26. $43 - 15 + 6 =$
27. $42 - 18 + 7 =$
28. $45 - 17 + 5 =$
29. $41 - 19 + 6 =$
30. $43 - 15 + 7 =$

Name: **Lesson 12-6 Adding and subtracting three numbers**

1. $43 - 18 + 6 =$
2. $42 - 14 + 7 =$
3. $47 - 18 + 8 =$
4. $44 - 19 + 6 =$
5. $41 - 15 + 7 =$
6. $45 - 17 + 8 =$
7. $41 - 12 + 6 =$
8. $43 - 17 + 7 =$
9. $43 - 19 + 8 =$
10. $43 - 15 + 6 =$
11. $44 - 15 + 7 =$
12. $41 - 13 + 8 =$
13. $46 - 17 + 6 =$
14. $41 - 14 + 7 =$
15. $45 - 16 + 8 =$
16. $46 - 18 + 6 =$
17. $44 - 16 + 7 =$
18. $45 - 19 + 8 =$
19. $46 - 19 + 6 =$
20. $44 - 18 + 7 =$
21. $43 - 14 + 8 =$
22. $41 - 17 + 6 =$
23. $42 - 15 + 7 =$
24. $45 - 18 + 8 =$
25. $42 - 16 + 6 =$
26. $44 - 17 + 7 =$
27. $41 - 18 + 8 =$
28. $43 - 16 + 6 =$
29. $42 - 13 + 7 =$
30. $43 - 16 + 8 =$

Name: **Lesson 12-7 Adding and subtracting three numbers**

---

1. $53 - 23 + 7 =$	2. $53 - 16 + 8 =$	3. $53 - 17 + 9 =$
4. $51 - 17 + 7 =$	5. $56 - 29 + 8 =$	6. $52 - 25 + 9 =$
7. $57 - 19 + 7 =$	8. $57 - 18 + 8 =$	9. $51 - 24 + 9 =$
10. $54 - 15 + 7 =$	11. $56 - 28 + 8 =$	12. $54 - 16 + 9 =$
13. $52 - 27 + 7 =$	14. $52 - 13 + 8 =$	15. $52 - 22 + 9 =$
16. $52 - 16 + 7 =$	17. $55 - 27 + 8 =$	18. $53 - 29 + 9 =$
19. $51 - 25 + 7 =$	20. $53 - 18 + 8 =$	21. $58 - 19 + 9 =$
22. $54 - 29 + 7 =$	23. $55 - 17 + 8 =$	24. $51 - 16 + 9 =$
25. $52 - 18 + 7 =$	26. $51 - 27 + 8 =$	27. $53 - 25 + 9 =$
28. $51 - 23 + 7 =$	29. $54 - 15 + 8 =$	30. $54 - 18 + 9 =$

Name: **Lesson 12-8 Adding and subtracting three numbers**

1. $53 - 26 + 8 =$
2. $54 - 25 + 9 =$
3. $53 - 17 + 10 =$
4. $55 - 18 + 8 =$
5. $51 - 12 + 9 =$
6. $51 - 29 + 10 =$
7. $51 - 17 + 8 =$
8. $56 - 29 + 9 =$
9. $52 - 23 + 10 =$
10. $55 - 26 + 8 =$
11. $52 - 25 + 9 =$
12. $57 - 28 + 10 =$
13. $53 - 19 + 8 =$
14. $54 - 16 + 9 =$
15. $56 - 27 + 10 =$
16. $55 - 17 + 8 =$
17. $52 - 27 + 9 =$
18. $58 - 19 + 10 =$
19. $55 - 29 + 8 =$
20. $51 - 13 + 9 =$
21. $54 - 18 + 10 =$
22. $51 - 16 + 8 =$
23. $52 - 28 + 9 =$
24. $53 - 23 + 10 =$
25. $52 - 28 + 8 =$
26. $54 - 18 + 9 =$
27. $55 - 27 + 10 =$
28. $54 - 17 + 8 =$
29. $51 - 15 + 9 =$
30. $56 - 18 + 10 =$

Name: **Lesson 12-9 Adding and subtracting three numbers**

---

1. $45 - 17 + 1 =$
2. $53 - 25 + 1 =$
3. $52 - 29 + 1 =$
4. $51 - 23 + 2 =$
5. $45 - 29 + 2 =$
6. $47 - 18 + 2 =$
7. $56 - 29 + 3 =$
8. $56 - 28 + 3 =$
9. $51 - 12 + 3 =$
10. $43 - 18 + 4 =$
11. $41 - 17 + 4 =$
12. $43 - 26 + 4 =$
13. $53 - 19 + 5 =$
14. $45 - 26 + 5 =$
15. $52 - 24 + 5 =$
16. $51 - 14 + 6 =$
17. $54 - 18 + 6 =$
18. $41 - 25 + 6 =$
19. $44 - 29 + 7 =$
20. $44 - 26 + 7 =$
21. $53 - 17 + 7 =$
22. $51 - 19 + 8 =$
23. $52 - 25 + 8 =$
24. $58 - 19 + 8 =$
25. $42 - 23 + 9 =$
26. $52 - 16 + 9 =$
27. $47 - 28 + 9 =$
28. $52 - 18 + 10 =$
29. $45 - 28 + 10 =$
30. $56 - 17 + 10 =$

Name: **Lesson 12-10 Adding and subtracting three numbers**

1. $54 - 16 + 1 =$
2. $41 - 29 + 1 =$
3. $51 - 19 + 1 =$
4. $55 - 19 + 2 =$
5. $53 - 17 + 2 =$
6. $42 - 19 + 2 =$
7. $47 - 28 + 3 =$
8. $54 - 15 + 3 =$
9. $43 - 24 + 3 =$
10. $55 - 27 + 4 =$
11. $46 - 18 + 4 =$
12. $52 - 15 + 4 =$
13. $41 - 13 + 5 =$
14. $53 - 29 + 5 =$
15. $45 - 19 + 5 =$
16. $43 - 25 + 6 =$
17. $45 - 26 + 6 =$
18. $52 - 27 + 6 =$
19. $52 - 17 + 7 =$
20. $52 - 14 + 7 =$
21. $43 - 17 + 7 =$
22. $51 - 23 + 8 =$
23. $41 - 22 + 8 =$
24. $51 - 25 + 8 =$
25. $42 - 17 + 9 =$
26. $52 - 18 + 9 =$
27. $47 - 19 + 9 =$
28. $54 - 26 + 10 =$
29. $42 - 15 + 10 =$
30. $56 - 18 + 10 =$

Name: **Lesson 1-1 Adding through 10**

1. 3 + 0 3	2. 1 + 4 5	3. 4 + 1 5	4. 0 + 2 2	5. 1 + 5 6
6. 2 + 3 5	7. 2 + 2 4	8. 0 + 5 5	9. 4 + 3 7	10. 8 + 0 8
11. 0 + 3 3	12. 2 + 5 7	13. 2 + 3 5	14. 1 + 6 7	15. 4 + 1 5
16. 6 + 3 9	17. 3 + 6 9	18. 5 + 4 9	19. 9 + 0 9	20. 2 + 6 8
21. 1 + 5 6	22. 5 + 0 5	23. 0 + 6 6	24. 2 + 7 9	25. 8 + 1 9
26. 3 + 3 6	27. 4 + 4 8	28. 1 + 7 8	29. 1 + 3 4	30. 4 + 2 6

Name: **Lesson 1-2 Adding through 11**

1. 1 + 1 2	2. 3 + 2 5	3. 3 + 1 4	4. 2 + 3 5	5. 3 + 3 6
6. 4 + 1 5	7. 1 + 4 5	8. 1 + 3 4	9. 6 + 1 7	10. 4 + 5 9
11. 2 + 1 3	12. 4 + 3 7	13. 0 + 5 5	14. 3 + 4 7	15. 2 + 2 4
16. 8 + 1 9	17. 5 + 4 9	18. 7 + 2 9	19. 2 + 6 8	20. 6 + 2 8
21. 3 + 6 9	22. 1 + 7 8	23. 2 + 4 6	24. 1 + 6 7	25. 2 + 7 9
26. 5 + 6 11	27. 7 + 3 10	28. 8 + 2 10	29. 6 + 5 11	30. 2 + 9 11

Name: **Lesson 1-3 Adding through 12**

1. 4 + 2 6	2. 1 + 4 5	3. 3 + 2 5	4. 2 + 4 6	5. 3 + 3 6
6. 2 + 3 5	7. 2 + 5 7	8. 1 + 4 5	9. 3 + 2 5	10. 2 + 7 9
11. 5 + 4 9	12. 3 + 4 7	13. 5 + 2 7	14. 5 + 1 6	15. 4 + 0 4
16. 6 + 3 9	17. 6 + 3 9	18. 3 + 0 3	19. 4 + 3 7	20. 5 + 4 9
21. 4 + 7 11	22. 8 + 2 10	23. 3 + 9 12	24. 2 + 9 11	25. 7 + 4 11
26. 9 + 3 12	27. 6 + 6 12	28. 7 + 4 11	29. 5 + 7 12	30. 6 + 4 10

Name: **Lesson 1-4 Adding through 13**

1. 5 + 2 7	2. 2 + 5 7	3. 5 + 3 8	4. 2 + 2 4	5. 3 + 1 4
6. 3 + 4 7	7. 8 + 0 8	8. 4 + 2 6	9. 3 + 5 8	10. 5 + 1 6
11. 4 + 4 8	12. 7 + 2 9	13. 4 + 3 7	14. 0 + 9 9	15. 2 + 6 8
16. 5 + 7 12	17. 2 + 4 6	18. 7 + 5 12	19. 9 + 2 11	20. 2 + 7 9
21. 8 + 4 12	22. 8 + 5 13	23. 9 + 4 13	24. 7 + 3 10	25. 9 + 3 12
26. 5 + 8 13	27. 4 + 9 13	28. 2 + 9 11	29. 3 + 9 12	30. 4 + 8 12

Name: Lesson 1-5 Adding through 14

1. 1 + 7 = 8	2. 5 + 2 = 7	3. 8 + 1 = 9	4. 6 + 3 = 9	5. 3 + 2 = 5
6. 6 + 2 = 8	7. 1 + 8 = 9	8. 2 + 3 = 5	9. 7 + 1 = 8	10. 4 + 5 = 9
11. 3 + 6 = 9	12. 5 + 3 = 8	13. 2 + 5 = 7	14. 2 + 6 = 8	15. 1 + 9 = 10
16. 5 + 5 = 10	17. 5 + 8 = 13	18. 2 + 8 = 10	19. 4 + 7 = 11	20. 8 + 2 = 10
21. 7 + 3 = 10	22. 8 + 6 = 14	23. 5 + 9 = 14	24. 3 + 7 = 10	25. 8 + 4 = 12
26. 9 + 2 = 11	27. 7 + 2 = 9	28. 7 + 7 = 14	29. 6 + 8 = 14	30. 9 + 5 = 14

Name: Lesson 1-6 Adding through 15

1. 3 + 4 = 7	2. 6 + 1 = 7	3. 3 + 2 = 5	4. 5 + 4 = 9	5. 4 + 4 = 8
6. 2 + 7 = 9	7. 2 + 6 = 8	8. 5 + 6 = 11	9. 7 + 3 = 10	10. 7 + 0 = 7
11. 3 + 5 = 8	12. 4 + 6 = 10	13. 8 + 5 = 13	14. 6 + 5 = 11	15. 6 + 6 = 12
16. 6 + 4 = 10	17. 5 + 7 = 12	18. 7 + 7 = 14	19. 7 + 8 = 15	20. 5 + 8 = 13
21. 8 + 7 = 15	22. 9 + 6 = 15	23. 5 + 9 = 14	24. 8 + 7 = 15	25. 7 + 5 = 12
26. 6 + 9 = 15	27. 6 + 7 = 13	28. 7 + 8 = 15	29. 7 + 6 = 13	30. 9 + 5 = 14

Name: Lesson 1-7 Adding through 16

1. 3 + 3 = 6	2. 5 + 3 = 8	3. 2 + 4 = 6	4. 1 + 7 = 8	5. 3 + 4 = 7
6. 7 + 2 = 9	7. 2 + 7 = 9	8. 5 + 2 = 7	9. 0 + 8 = 8	10. 8 + 1 = 9
11. 4 + 0 = 4	12. 4 + 6 = 10	13. 9 + 4 = 13	14. 5 + 6 = 11	15. 7 + 5 = 12
16. 6 + 5 = 11	17. 6 + 8 = 14	18. 5 + 7 = 12	19. 5 + 8 = 13	20. 8 + 6 = 14
21. 7 + 7 = 14	22. 7 + 6 = 13	23. 7 + 9 = 16	24. 7 + 8 = 15	25. 5 + 7 = 12
26. 4 + 9 = 13	27. 8 + 8 = 16	28. 6 + 7 = 13	29. 9 + 7 = 16	30. 8 + 7 = 15

Name: Lesson 1-8 Adding through 17

1. 5 + 3 = 8	2. 2 + 5 = 7	3. 3 + 5 = 8	4. 3 + 6 = 9	5. 8 + 1 = 9
6. 6 + 7 = 13	7. 6 + 3 = 9	8. 7 + 5 = 12	9. 5 + 2 = 7	10. 4 + 8 = 12
11. 3 + 8 = 11	12. 7 + 6 = 13	13. 6 + 7 = 13	14. 5 + 7 = 12	15. 6 + 6 = 12
16. 6 + 9 = 15	17. 8 + 4 = 12	18. 9 + 6 = 15	19. 7 + 7 = 14	20. 9 + 4 = 13
21. 9 + 7 = 16	22. 7 + 8 = 15	23. 7 + 9 = 16	24. 5 + 8 = 13	25. 8 + 5 = 13
26. 8 + 9 = 17	27. 8 + 8 = 16	28. 9 + 8 = 17	29. 8 + 7 = 15	30. 7 + 6 = 13

Name: **Lesson 1-9 Adding through 18**

1. $\begin{array}{r} 3 \\ +\ 7 \\ \hline 10 \end{array}$	2. $\begin{array}{r} 4 \\ +\ 6 \\ \hline 10 \end{array}$	3. $\begin{array}{r} 8 \\ +\ 2 \\ \hline 10 \end{array}$	4. $\begin{array}{r} 1 \\ +\ 9 \\ \hline 10 \end{array}$	5. $\begin{array}{r} 9 \\ +\ 1 \\ \hline 10 \end{array}$
6. $\begin{array}{r} 6 \\ +\ 4 \\ \hline 10 \end{array}$	7. $\begin{array}{r} 5 \\ +\ 7 \\ \hline 12 \end{array}$	8. $\begin{array}{r} 6 \\ +\ 4 \\ \hline 10 \end{array}$	9. $\begin{array}{r} 7 \\ +\ 4 \\ \hline 11 \end{array}$	10. $\begin{array}{r} 5 \\ +\ 6 \\ \hline 11 \end{array}$
11. $\begin{array}{r} 7 \\ +\ 6 \\ \hline 13 \end{array}$	12. $\begin{array}{r} 4 \\ +\ 7 \\ \hline 11 \end{array}$	13. $\begin{array}{r} 7 \\ +\ 5 \\ \hline 12 \end{array}$	14. $\begin{array}{r} 5 \\ +\ 8 \\ \hline 13 \end{array}$	15. $\begin{array}{r} 6 \\ +\ 8 \\ \hline 14 \end{array}$
16. $\begin{array}{r} 6 \\ +\ 7 \\ \hline 13 \end{array}$	17. $\begin{array}{r} 8 \\ +\ 7 \\ \hline 15 \end{array}$	18. $\begin{array}{r} 8 \\ +\ 6 \\ \hline 14 \end{array}$	19. $\begin{array}{r} 7 \\ +\ 8 \\ \hline 15 \end{array}$	20. $\begin{array}{r} 7 \\ +\ 7 \\ \hline 14 \end{array}$
21. $\begin{array}{r} 9 \\ +\ 7 \\ \hline 16 \end{array}$	22. $\begin{array}{r} 9 \\ +\ 6 \\ \hline 15 \end{array}$	23. $\begin{array}{r} 5 \\ +\ 9 \\ \hline 14 \end{array}$	24. $\begin{array}{r} 9 \\ +\ 8 \\ \hline 17 \end{array}$	25. $\begin{array}{r} 9 \\ +\ 5 \\ \hline 14 \end{array}$
26. $\begin{array}{r} 9 \\ +\ 9 \\ \hline 18 \end{array}$	27. $\begin{array}{r} 8 \\ +\ 9 \\ \hline 17 \end{array}$	28. $\begin{array}{r} 8 \\ +\ 8 \\ \hline 16 \end{array}$	29. $\begin{array}{r} 7 \\ +\ 9 \\ \hline 16 \end{array}$	30. $\begin{array}{r} 6 \\ +\ 9 \\ \hline 15 \end{array}$

Name: **Lesson 1-10 Adding through 19**

1. $\begin{array}{r} 2 \\ +\ 9 \\ \hline 11 \end{array}$	2. $\begin{array}{r} 5 \\ +\ 5 \\ \hline 10 \end{array}$	3. $\begin{array}{r} 9 \\ +\ 2 \\ \hline 11 \end{array}$	4. $\begin{array}{r} 6 \\ +\ 5 \\ \hline 11 \end{array}$	5. $\begin{array}{r} 1 \\ +\ 9 \\ \hline 10 \end{array}$
6. $\begin{array}{r} 3 \\ +\ 10 \\ \hline 13 \end{array}$	7. $\begin{array}{r} 5 \\ +\ 6 \\ \hline 11 \end{array}$	8. $\begin{array}{r} 6 \\ +\ 8 \\ \hline 14 \end{array}$	9. $\begin{array}{r} 8 \\ +\ 6 \\ \hline 14 \end{array}$	10. $\begin{array}{r} 6 \\ +\ 6 \\ \hline 12 \end{array}$
11. $\begin{array}{r} 3 \\ +\ 8 \\ \hline 11 \end{array}$	12. $\begin{array}{r} 6 \\ +\ 9 \\ \hline 15 \end{array}$	13. $\begin{array}{r} 7 \\ +\ 7 \\ \hline 14 \end{array}$	14. $\begin{array}{r} 8 \\ +\ 7 \\ \hline 15 \end{array}$	15. $\begin{array}{r} 7 \\ +\ 8 \\ \hline 15 \end{array}$
16. $\begin{array}{r} 4 \\ +\ 7 \\ \hline 11 \end{array}$	17. $\begin{array}{r} 7 \\ +\ 10 \\ \hline 17 \end{array}$	18. $\begin{array}{r} 5 \\ +\ 8 \\ \hline 13 \end{array}$	19. $\begin{array}{r} 9 \\ +\ 6 \\ \hline 15 \end{array}$	20. $\begin{array}{r} 8 \\ +\ 5 \\ \hline 13 \end{array}$
21. $\begin{array}{r} 8 \\ +\ 8 \\ \hline 16 \end{array}$	22. $\begin{array}{r} 9 \\ +\ 7 \\ \hline 16 \end{array}$	23. $\begin{array}{r} 10 \\ +\ 8 \\ \hline 18 \end{array}$	24. $\begin{array}{r} 10 \\ +\ 7 \\ \hline 17 \end{array}$	25. $\begin{array}{r} 9 \\ +\ 8 \\ \hline 17 \end{array}$
26. $\begin{array}{r} 10 \\ +\ 9 \\ \hline 19 \end{array}$	27. $\begin{array}{r} 8 \\ +\ 9 \\ \hline 17 \end{array}$	28. $\begin{array}{r} 9 \\ +\ 10 \\ \hline 19 \end{array}$	29. $\begin{array}{r} 7 \\ +\ 9 \\ \hline 16 \end{array}$	30. $\begin{array}{r} 8 \\ +\ 10 \\ \hline 18 \end{array}$

Name: **Lesson 2-1 Subtracting two numbers**

1. $\begin{array}{r} 3 \\ -\ 1 \\ \hline 2 \end{array}$	2. $\begin{array}{r} 4 \\ -\ 2 \\ \hline 2 \end{array}$	3. $\begin{array}{r} 4 \\ -\ 1 \\ \hline 3 \end{array}$	4. $\begin{array}{r} 5 \\ -\ 3 \\ \hline 2 \end{array}$	5. $\begin{array}{r} 3 \\ -\ 0 \\ \hline 3 \end{array}$
6. $\begin{array}{r} 5 \\ -\ 2 \\ \hline 3 \end{array}$	7. $\begin{array}{r} 5 \\ -\ 4 \\ \hline 1 \end{array}$	8. $\begin{array}{r} 5 \\ -\ 1 \\ \hline 4 \end{array}$	9. $\begin{array}{r} 6 \\ -\ 4 \\ \hline 2 \end{array}$	10. $\begin{array}{r} 5 \\ -\ 5 \\ \hline 0 \end{array}$
11. $\begin{array}{r} 6 \\ -\ 3 \\ \hline 3 \end{array}$	12. $\begin{array}{r} 6 \\ -\ 2 \\ \hline 4 \end{array}$	13. $\begin{array}{r} 7 \\ -\ 4 \\ \hline 3 \end{array}$	14. $\begin{array}{r} 7 \\ -\ 3 \\ \hline 4 \end{array}$	15. $\begin{array}{r} 6 \\ -\ 5 \\ \hline 1 \end{array}$
16. $\begin{array}{r} 7 \\ -\ 2 \\ \hline 5 \end{array}$	17. $\begin{array}{r} 7 \\ -\ 1 \\ \hline 6 \end{array}$	18. $\begin{array}{r} 8 \\ -\ 1 \\ \hline 7 \end{array}$	19. $\begin{array}{r} 8 \\ -\ 5 \\ \hline 3 \end{array}$	20. $\begin{array}{r} 7 \\ -\ 6 \\ \hline 1 \end{array}$
21. $\begin{array}{r} 9 \\ -\ 5 \\ \hline 4 \end{array}$	22. $\begin{array}{r} 8 \\ -\ 7 \\ \hline 1 \end{array}$	23. $\begin{array}{r} 8 \\ -\ 8 \\ \hline 0 \end{array}$	24. $\begin{array}{r} 9 \\ -\ 3 \\ \hline 6 \end{array}$	25. $\begin{array}{r} 8 \\ -\ 4 \\ \hline 4 \end{array}$
26. $\begin{array}{r} 8 \\ -\ 6 \\ \hline 2 \end{array}$	27. $\begin{array}{r} 9 \\ -\ 4 \\ \hline 5 \end{array}$	28. $\begin{array}{r} 9 \\ -\ 8 \\ \hline 1 \end{array}$	29. $\begin{array}{r} 9 \\ -\ 9 \\ \hline 0 \end{array}$	30. $\begin{array}{r} 9 \\ -\ 2 \\ \hline 7 \end{array}$

Name: **Lesson 2-2 Subtracting two numbers**

1. $\begin{array}{r} 6 \\ -\ 4 \\ \hline 2 \end{array}$	2. $\begin{array}{r} 9 \\ -\ 8 \\ \hline 1 \end{array}$	3. $\begin{array}{r} 6 \\ -\ 5 \\ \hline 1 \end{array}$	4. $\begin{array}{r} 9 \\ -\ 4 \\ \hline 5 \end{array}$	5. $\begin{array}{r} 8 \\ -\ 7 \\ \hline 1 \end{array}$
6. $\begin{array}{r} 7 \\ -\ 6 \\ \hline 1 \end{array}$	7. $\begin{array}{r} 5 \\ -\ 2 \\ \hline 3 \end{array}$	8. $\begin{array}{r} 5 \\ -\ 1 \\ \hline 4 \end{array}$	9. $\begin{array}{r} 7 \\ -\ 5 \\ \hline 2 \end{array}$	10. $\begin{array}{r} 8 \\ -\ 4 \\ \hline 4 \end{array}$
11. $\begin{array}{r} 8 \\ -\ 1 \\ \hline 7 \end{array}$	12. $\begin{array}{r} 7 \\ -\ 2 \\ \hline 5 \end{array}$	13. $\begin{array}{r} 8 \\ -\ 0 \\ \hline 8 \end{array}$	14. $\begin{array}{r} 9 \\ -\ 3 \\ \hline 6 \end{array}$	15. $\begin{array}{r} 9 \\ -\ 7 \\ \hline 2 \end{array}$
16. $\begin{array}{r} 6 \\ -\ 6 \\ \hline 0 \end{array}$	17. $\begin{array}{r} 9 \\ -\ 4 \\ \hline 5 \end{array}$	18. $\begin{array}{r} 5 \\ -\ 3 \\ \hline 2 \end{array}$	19. $\begin{array}{r} 5 \\ -\ 4 \\ \hline 1 \end{array}$	20. $\begin{array}{r} 9 \\ -\ 2 \\ \hline 7 \end{array}$
21. $\begin{array}{r} 5 \\ -\ 5 \\ \hline 0 \end{array}$	22. $\begin{array}{r} 8 \\ -\ 2 \\ \hline 6 \end{array}$	23. $\begin{array}{r} 8 \\ -\ 7 \\ \hline 1 \end{array}$	24. $\begin{array}{r} 6 \\ -\ 3 \\ \hline 3 \end{array}$	25. $\begin{array}{r} 7 \\ -\ 6 \\ \hline 1 \end{array}$
26. $\begin{array}{r} 7 \\ -\ 0 \\ \hline 7 \end{array}$	27. $\begin{array}{r} 5 \\ -\ 3 \\ \hline 2 \end{array}$	28. $\begin{array}{r} 6 \\ -\ 2 \\ \hline 4 \end{array}$	29. $\begin{array}{r} 7 \\ -\ 4 \\ \hline 3 \end{array}$	30. $\begin{array}{r} 9 \\ -\ 3 \\ \hline 6 \end{array}$

Name: **Lesson 2-3 Subtracting two numbers**

1. $\begin{array}{r} 8 \\ -\ 1 \\ \hline 7 \end{array}$	2. $\begin{array}{r} 5 \\ -\ 3 \\ \hline 2 \end{array}$	3. $\begin{array}{r} 6 \\ -\ 2 \\ \hline 4 \end{array}$	4. $\begin{array}{r} 5 \\ -\ 2 \\ \hline 3 \end{array}$	5. $\begin{array}{r} 8 \\ -\ 3 \\ \hline 5 \end{array}$
6. $\begin{array}{r} 4 \\ -\ 0 \\ \hline 4 \end{array}$	7. $\begin{array}{r} 3 \\ -\ 1 \\ \hline 2 \end{array}$	8. $\begin{array}{r} 4 \\ -\ 2 \\ \hline 2 \end{array}$	9. $\begin{array}{r} 8 \\ -\ 7 \\ \hline 1 \end{array}$	10. $\begin{array}{r} 6 \\ -\ 5 \\ \hline 1 \end{array}$
11. $\begin{array}{r} 7 \\ -\ 3 \\ \hline 4 \end{array}$	12. $\begin{array}{r} 8 \\ -\ 6 \\ \hline 2 \end{array}$	13. $\begin{array}{r} 7 \\ -\ 4 \\ \hline 3 \end{array}$	14. $\begin{array}{r} 9 \\ -\ 9 \\ \hline 0 \end{array}$	15. $\begin{array}{r} 7 \\ -\ 0 \\ \hline 7 \end{array}$
16. $\begin{array}{r} 5 \\ -\ 4 \\ \hline 1 \end{array}$	17. $\begin{array}{r} 5 \\ -\ 5 \\ \hline 0 \end{array}$	18. $\begin{array}{r} 6 \\ -\ 3 \\ \hline 3 \end{array}$	19. $\begin{array}{r} 3 \\ -\ 2 \\ \hline 1 \end{array}$	20. $\begin{array}{r} 7 \\ -\ 2 \\ \hline 5 \end{array}$
21. $\begin{array}{r} 6 \\ -\ 0 \\ \hline 6 \end{array}$	22. $\begin{array}{r} 7 \\ -\ 5 \\ \hline 2 \end{array}$	23. $\begin{array}{r} 8 \\ -\ 5 \\ \hline 3 \end{array}$	24. $\begin{array}{r} 5 \\ -\ 1 \\ \hline 4 \end{array}$	25. $\begin{array}{r} 4 \\ -\ 4 \\ \hline 0 \end{array}$
26. $\begin{array}{r} 7 \\ -\ 7 \\ \hline 0 \end{array}$	27. $\begin{array}{r} 6 \\ -\ 1 \\ \hline 5 \end{array}$	28. $\begin{array}{r} 7 \\ -\ 1 \\ \hline 6 \end{array}$	29. $\begin{array}{r} 8 \\ -\ 2 \\ \hline 6 \end{array}$	30. $\begin{array}{r} 6 \\ -\ 0 \\ \hline 6 \end{array}$

Name: **Lesson 2-4 Subtracting two numbers**

1. $\begin{array}{r} 3 \\ -\ 0 \\ \hline 3 \end{array}$	2. $\begin{array}{r} 8 \\ -\ 2 \\ \hline 6 \end{array}$	3. $\begin{array}{r} 5 \\ -\ 1 \\ \hline 4 \end{array}$	4. $\begin{array}{r} 6 \\ -\ 3 \\ \hline 3 \end{array}$	5. $\begin{array}{r} 8 \\ -\ 5 \\ \hline 3 \end{array}$
6. $\begin{array}{r} 7 \\ -\ 4 \\ \hline 3 \end{array}$	7. $\begin{array}{r} 2 \\ -\ 0 \\ \hline 2 \end{array}$	8. $\begin{array}{r} 4 \\ -\ 3 \\ \hline 1 \end{array}$	9. $\begin{array}{r} 7 \\ -\ 0 \\ \hline 7 \end{array}$	10. $\begin{array}{r} 3 \\ -\ 2 \\ \hline 1 \end{array}$
11. $\begin{array}{r} 5 \\ -\ 3 \\ \hline 2 \end{array}$	12. $\begin{array}{r} 6 \\ -\ 0 \\ \hline 6 \end{array}$	13. $\begin{array}{r} 2 \\ -\ 1 \\ \hline 1 \end{array}$	14. $\begin{array}{r} 8 \\ -\ 4 \\ \hline 4 \end{array}$	15. $\begin{array}{r} 6 \\ -\ 2 \\ \hline 4 \end{array}$
16. $\begin{array}{r} 7 \\ -\ 6 \\ \hline 1 \end{array}$	17. $\begin{array}{r} 7 \\ -\ 1 \\ \hline 6 \end{array}$	18. $\begin{array}{r} 8 \\ -\ 6 \\ \hline 2 \end{array}$	19. $\begin{array}{r} 4 \\ -\ 0 \\ \hline 4 \end{array}$	20. $\begin{array}{r} 7 \\ -\ 5 \\ \hline 2 \end{array}$
21. $\begin{array}{r} 4 \\ -\ 1 \\ \hline 3 \end{array}$	22. $\begin{array}{r} 4 \\ -\ 4 \\ \hline 0 \end{array}$	23. $\begin{array}{r} 8 \\ -\ 3 \\ \hline 5 \end{array}$	24. $\begin{array}{r} 8 \\ -\ 1 \\ \hline 7 \end{array}$	25. $\begin{array}{r} 5 \\ -\ 4 \\ \hline 1 \end{array}$
26. $\begin{array}{r} 6 \\ -\ 4 \\ \hline 2 \end{array}$	27. $\begin{array}{r} 7 \\ -\ 2 \\ \hline 5 \end{array}$	28. $\begin{array}{r} 7 \\ -\ 3 \\ \hline 4 \end{array}$	29. $\begin{array}{r} 8 \\ -\ 7 \\ \hline 1 \end{array}$	30. $\begin{array}{r} 8 \\ -\ 8 \\ \hline 0 \end{array}$

Name: **Lesson 2-5 Subtracting two numbers**

1. $\begin{array}{r} 9 \\ -\ 4 \\ \hline 5 \end{array}$	2. $\begin{array}{r} 8 \\ -\ 2 \\ \hline 6 \end{array}$	3. $\begin{array}{r} 6 \\ -\ 2 \\ \hline 4 \end{array}$	4. $\begin{array}{r} 6 \\ -\ 3 \\ \hline 3 \end{array}$	5. $\begin{array}{r} 9 \\ -\ 3 \\ \hline 6 \end{array}$
6. $\begin{array}{r} 9 \\ -\ 0 \\ \hline 9 \end{array}$	7. $\begin{array}{r} 7 \\ -\ 3 \\ \hline 4 \end{array}$	8. $\begin{array}{r} 8 \\ -\ 4 \\ \hline 4 \end{array}$	9. $\begin{array}{r} 8 \\ -\ 5 \\ \hline 3 \end{array}$	10. $\begin{array}{r} 9 \\ -\ 8 \\ \hline 1 \end{array}$
11. $\begin{array}{r} 3 \\ -\ 2 \\ \hline 1 \end{array}$	12. $\begin{array}{r} 9 \\ -\ 9 \\ \hline 0 \end{array}$	13. $\begin{array}{r} 6 \\ -\ 5 \\ \hline 1 \end{array}$	14. $\begin{array}{r} 6 \\ -\ 3 \\ \hline 3 \end{array}$	15. $\begin{array}{r} 8 \\ -\ 6 \\ \hline 2 \end{array}$
16. $\begin{array}{r} 6 \\ -\ 4 \\ \hline 2 \end{array}$	17. $\begin{array}{r} 7 \\ -\ 2 \\ \hline 5 \end{array}$	18. $\begin{array}{r} 5 \\ -\ 2 \\ \hline 3 \end{array}$	19. $\begin{array}{r} 8 \\ -\ 5 \\ \hline 3 \end{array}$	20. $\begin{array}{r} 5 \\ -\ 5 \\ \hline 0 \end{array}$
21. $\begin{array}{r} 8 \\ -\ 7 \\ \hline 1 \end{array}$	22. $\begin{array}{r} 7 \\ -\ 7 \\ \hline 0 \end{array}$	23. $\begin{array}{r} 7 \\ -\ 5 \\ \hline 2 \end{array}$	24. $\begin{array}{r} 8 \\ -\ 3 \\ \hline 5 \end{array}$	25. $\begin{array}{r} 4 \\ -\ 2 \\ \hline 2 \end{array}$
26. $\begin{array}{r} 8 \\ -\ 0 \\ \hline 8 \end{array}$	27. $\begin{array}{r} 5 \\ -\ 3 \\ \hline 2 \end{array}$	28. $\begin{array}{r} 9 \\ -\ 5 \\ \hline 4 \end{array}$	29. $\begin{array}{r} 4 \\ -\ 3 \\ \hline 1 \end{array}$	30. $\begin{array}{r} 7 \\ -\ 6 \\ \hline 1 \end{array}$

Name: **Lesson 2-6 Subtracting two numbers**

1. $\begin{array}{r} 8 \\ -\ 6 \\ \hline 2 \end{array}$	2. $\begin{array}{r} 5 \\ -\ 2 \\ \hline 3 \end{array}$	3. $\begin{array}{r} 6 \\ -\ 4 \\ \hline 2 \end{array}$	4. $\begin{array}{r} 7 \\ -\ 5 \\ \hline 2 \end{array}$	5. $\begin{array}{r} 3 \\ -\ 3 \\ \hline 0 \end{array}$
6. $\begin{array}{r} 4 \\ -\ 2 \\ \hline 2 \end{array}$	7. $\begin{array}{r} 8 \\ -\ 8 \\ \hline 0 \end{array}$	8. $\begin{array}{r} 4 \\ -\ 0 \\ \hline 4 \end{array}$	9. $\begin{array}{r} 6 \\ -\ 4 \\ \hline 2 \end{array}$	10. $\begin{array}{r} 9 \\ -\ 8 \\ \hline 1 \end{array}$
11. $\begin{array}{r} 9 \\ -\ 9 \\ \hline 0 \end{array}$	12. $\begin{array}{r} 5 \\ -\ 0 \\ \hline 5 \end{array}$	13. $\begin{array}{r} 9 \\ -\ 7 \\ \hline 2 \end{array}$	14. $\begin{array}{r} 8 \\ -\ 5 \\ \hline 3 \end{array}$	15. $\begin{array}{r} 3 \\ -\ 2 \\ \hline 1 \end{array}$
16. $\begin{array}{r} 5 \\ -\ 4 \\ \hline 1 \end{array}$	17. $\begin{array}{r} 8 \\ -\ 3 \\ \hline 5 \end{array}$	18. $\begin{array}{r} 8 \\ -\ 2 \\ \hline 6 \end{array}$	19. $\begin{array}{r} 7 \\ -\ 3 \\ \hline 4 \end{array}$	20. $\begin{array}{r} 7 \\ -\ 4 \\ \hline 3 \end{array}$
21. $\begin{array}{r} 9 \\ -\ 4 \\ \hline 5 \end{array}$	22. $\begin{array}{r} 4 \\ -\ 3 \\ \hline 1 \end{array}$	23. $\begin{array}{r} 7 \\ -\ 6 \\ \hline 1 \end{array}$	24. $\begin{array}{r} 9 \\ -\ 2 \\ \hline 7 \end{array}$	25. $\begin{array}{r} 5 \\ -\ 3 \\ \hline 2 \end{array}$
26. $\begin{array}{r} 9 \\ -\ 0 \\ \hline 9 \end{array}$	27. $\begin{array}{r} 9 \\ -\ 3 \\ \hline 6 \end{array}$	28. $\begin{array}{r} 9 \\ -\ 5 \\ \hline 4 \end{array}$	29. $\begin{array}{r} 9 \\ -\ 6 \\ \hline 3 \end{array}$	30. $\begin{array}{r} 6 \\ -\ 2 \\ \hline 4 \end{array}$

Name:

Lesson 2-7 Subtracting from 10, 11, and 12

1. 10 − 1 = 9	2. 10 − 2 = 8	3. 10 − 4 = 6	4. 10 − 5 = 5	5. 10 − 8 = 2
6. 10 − 3 = 7	7. 11 − 3 = 8	8. 11 − 5 = 6	9. 11 − 6 = 5	10. 11 − 7 = 4
11. 10 − 6 = 4	12. 10 − 7 = 3	13. 11 − 9 = 2	14. 11 − 10 = 1	15. 12 − 10 = 2
16. 10 − 8 = 2	17. 10 − 9 = 1	18. 12 − 9 = 3	19. 12 − 8 = 4	20. 12 − 6 = 6
21. 11 − 4 = 7	22. 12 − 11 = 1	23. 12 − 4 = 8	24. 12 − 11 = 1	25. 12 − 5 = 7
26. 12 − 10 = 2	27. 12 − 9 = 3	28. 12 − 7 = 5	29. 12 − 6 = 6	30. 12 − 3 = 9

Name:

Lesson 2-8 Subtracting from 13, 14, and 15

1. 13 − 2 = 11	2. 13 − 4 = 9	3. 13 − 5 = 8	4. 13 − 9 = 4	5. 13 − 10 = 3
6. 13 − 0 = 13	7. 13 − 6 = 7	8. 13 − 8 = 5	9. 14 − 11 = 3	10. 14 − 4 = 10
11. 14 − 3 = 11	12. 14 − 0 = 14	13. 14 − 12 = 2	14. 14 − 8 = 6	15. 14 − 10 = 4
16. 14 − 14 = 0	17. 15 − 2 = 13	18. 14 − 9 = 5	19. 15 − 9 = 6	20. 14 − 7 = 7
21. 15 − 0 = 15	22. 15 − 14 = 1	23. 15 − 8 = 7	24. 15 − 6 = 9	25. 15 − 10 = 5
26. 15 − 13 = 2	27. 15 − 5 = 10	28. 15 − 7 = 8	29. 15 − 12 = 3	30. 15 − 4 = 11

Name:

Lesson 2-9 Subtracting from 16, 17, and 18

1. 16 − 1 = 15	2. 16 − 2 = 14	3. 16 − 3 = 13	4. 16 − 5 = 11	5. 16 − 15 = 1
6. 16 − 13 = 3	7. 16 − 11 = 5	8. 16 − 10 = 6	9. 16 − 9 = 7	10. 16 − 7 = 9
11. 17 − 17 = 0	12. 16 − 8 = 8	13. 16 − 7 = 9	14. 16 − 6 = 10	15. 17 − 3 = 14
16. 17 − 12 = 5	17. 17 − 5 = 12	18. 17 − 14 = 3	19. 17 − 11 = 6	20. 17 − 9 = 8
21. 18 − 18 = 0	22. 17 − 10 = 7	23. 17 − 8 = 9	24. 18 − 10 = 8	25. 18 − 12 = 6
26. 18 − 2 = 16	27. 18 − 9 = 9	28. 18 − 8 = 10	29. 18 − 7 = 11	30. 18 − 6 = 12

Name:

Lesson 2-10 Subtracting from 17, 18, and 19

1. 17 − 15 = 2	2. 17 − 14 = 3	3. 17 − 13 = 4	4. 17 − 12 = 5	5. 17 − 11 = 6
6. 17 − 2 = 15	7. 17 − 3 = 14	8. 17 − 4 = 13	9. 17 − 5 = 12	10. 18 − 8 = 10
11. 18 − 3 = 15	12. 18 − 2 = 16	13. 18 − 5 = 13	14. 18 − 7 = 11	15. 18 − 10 = 8
16. 18 − 15 = 3	17. 19 − 8 = 11	18. 18 − 13 = 5	19. 18 − 11 = 7	20. 18 − 12 = 6
21. 19 − 10 = 9	22. 19 − 11 = 8	23. 19 − 4 = 15	24. 19 − 7 = 12	25. 18 − 6 = 12
26. 19 − 9 = 10	27. 19 − 13 = 6	28. 19 − 15 = 4	29. 19 − 12 = 7	30. 19 − 8 = 11

Name: Lesson 3-1 Adding three numbers

1. $3+6+1=10$	2. $5+5+2=12$	3. $1+2+3=6$
4. $2+2+1=5$	5. $3+7+2=12$	6. $4+5+3=12$
7. $4+5+1=10$	8. $7+4+2=13$	9. $1+4+3=8$
10. $8+2+1=11$	11. $4+3+2=9$	12. $5+2+3=10$
13. $6+4+1=11$	14. $2+6+2=10$	15. $2+7+3=12$
16. $2+7+1=10$	17. $4+4+2=10$	18. $5+4+3=12$
19. $5+4+1=10$	20. $2+8+2=12$	21. $6+1+3=10$
22. $2+9+1=12$	23. $4+6+2=12$	24. $2+3+3=8$
25. $3+2+1=6$	26. $5+3+2=10$	27. $5+1+3=9$
28. $9+2+1=12$	29. $4+2+2=8$	30. $7+2+3=12$

Name: Lesson 3-2 Adding three numbers

1. $2+2+2=6$	2. $3+6+3=12$	3. $4+3+4=11$
4. $6+4+2=12$	5. $6+5+3=14$	6. $2+4+4=10$
7. $7+3+2=12$	8. $8+2+3=13$	9. $2+6+4=12$
10. $7+2+2=11$	11. $4+6+3=13$	12. $5+5+4=14$
13. $6+3+2=11$	14. $2+8+3=13$	15. $7+4+4=15$
16. $8+3+2=13$	17. $3+5+3=11$	18. $4+4+4=12$
19. $5+2+2=9$	20. $5+4+3=12$	21. $3+3+4=10$
22. $3+8+2=13$	23. $2+9+3=14$	24. $2+3+4=9$
25. $3+7+2=12$	26. $4+7+3=14$	27. $5+3+4=12$
28. $5+6+2=13$	29. $2+7+3=12$	30. $4+5+4=13$

Name: Lesson 3-3 Adding three numbers

1. $7+2+3=12$	2. $5+5+4=14$	3. $2+9+5=16$
4. $3+5+3=11$	5. $8+2+4=14$	6. $4+5+5=14$
7. $2+4+3=9$	8. $3+6+4=13$	9. $2+3+5=10$
10. $4+3+3=10$	11. $5+4+4=13$	12. $5+2+5=12$
13. $2+6+3=11$	14. $8+3+4=15$	15. $6+5+5=16$
16. $4+7+3=14$	17. $5+3+4=12$	18. $2+2+5=9$
19. $2+8+3=13$	20. $7+3+4=14$	21. $3+4+5=12$
22. $4+6+3=13$	23. $2+5+4=11$	24. $3+8+5=16$
25. $3+2+3=8$	26. $6+4+4=14$	27. $2+7+5=14$
28. $6+3+3=12$	29. $9+2+4=15$	30. $4+4+5=13$

Name: Lesson 3-4 Adding three numbers

1. $8+2+4=14$	2. $7+2+5=14$	3. $3+5+6=14$
4. $3+2+4=9$	5. $2+6+5=13$	6. $5+2+6=13$
7. $9+2+4=15$	8. $3+4+5=12$	9. $2+2+6=10$
10. $7+4+4=15$	11. $5+6+5=16$	12. $4+2+6=12$
13. $5+4+4=13$	14. $2+3+5=10$	15. $6+3+6=15$
16. $2+8+4=14$	17. $3+7+5=15$	18. $2+9+6=17$
19. $2+5+4=11$	20. $8+3+5=16$	21. $6+4+6=16$
22. $7+3+4=14$	23. $4+5+5=14$	24. $5+5+6=16$
25. $6+5+4=14$	26. $3+8+5=16$	27. $2+4+6=12$
28. $6+3+4=13$	29. $2+7+5=14$	30. $4+7+6=17$

Name: Lesson 3-5 Adding three numbers

1. $9+2+5=16$ 2. $6+3+6=15$ 3. $5+2+7=14$

4. $5+3+5=13$ 5. $4+7+6=17$ 6. $3+8+7=18$

7. $2+6+5=13$ 8. $2+3+6=11$ 9. $2+8+7=17$

10. $5+4+5=14$ 11. $4+2+6=12$ 12. $3+3+7=13$

13. $2+7+5=14$ 14. $3+4+6=13$ 15. $7+3+7=17$

16. $3+5+5=13$ 17. $4+6+6=16$ 18. $2+4+7=13$

19. $4+4+5=13$ 20. $2+9+6=17$ 21. $6+4+7=17$

22. $5+1+5=11$ 23. $3+7+6=16$ 24. $7+2+7=16$

25. $3+2+5=10$ 26. $4+5+6=15$ 27. $2+5+7=14$

28. $8+2+5=15$ 29. $8+3+6=17$ 30. $7+4+7=18$

Name: Lesson 3-6 Adding three numbers

1. $4+3+6=13$ 2. $2+9+7=18$ 3. $4+2+8=14$

4. $6+2+6=14$ 5. $3+3+7=13$ 6. $2+7+8=17$

7. $5+4+6=15$ 8. $7+3+7=17$ 9. $4+4+8=16$

10. $3+6+6=15$ 11. $2+2+7=11$ 12. $5+2+8=15$

13. $6+5+6=17$ 14. $8+2+7=17$ 15. $4+7+8=19$

16. $5+5+6=16$ 17. $6+3+7=16$ 18. $4+6+8=18$

19. $8+3+6=17$ 20. $3+5+7=15$ 21. $2+4+8=14$

22. $7+4+6=17$ 23. $6+4+7=17$ 24. $9+2+8=19$

25. $3+2+6=11$ 26. $2+6+7=15$ 27. $3+8+8=19$

28. $8+4+6=18$ 29. $3+7+7=17$ 30. $5+6+8=19$

Name: Lesson 3-7 Adding three numbers

1. $9+2+7=18$ 2. $5+4+8=17$ 3. $2+2+9=13$

4. $2+6+7=15$ 5. $8+2+8=18$ 6. $4+3+9=16$

7. $3+3+7=13$ 8. $2+3+8=13$ 9. $6+2+9=17$

10. $2+5+7=14$ 11. $7+2+8=17$ 12. $5+6+9=20$

13. $7+4+7=18$ 14. $2+4+8=14$ 15. $8+3+9=20$

16. $2+7+7=16$ 17. $3+8+8=14$ 18. $4+4+9=17$

19. $6+3+7=16$ 20. $6+4+8=18$ 21. $3+6+9=18$

22. $5+3+7=15$ 23. $5+5+8=18$ 24. $2+8+9=19$

25. $3+7+7=17$ 26. $3+2+8=13$ 27. $4+6+9=19$

28. $5+2+7=14$ 29. $4+5+8=17$ 30. $7+3+9=19$

Name: Lesson 3-8 Adding three numbers

1. $5+2+8=15$ 2. $7+4+9=20$ 3. $2+6+10=18$

4. $6+5+8=19$ 5. $3+5+9=17$ 6. $4+5+10=19$

7. $7+2+8=17$ 8. $6+3+9=18$ 9. $5+5+10=20$

10. $4+3+8=15$ 11. $2+4+9=15$ 12. $3+2+10=15$

13. $8+2+8=18$ 14. $4+6+9=19$ 15. $9+2+10=21$

16. $3+7+8=18$ 17. $3+8+9=20$ 18. $4+7+10=21$

19. $6+4+8=18$ 20. $3+4+9=16$ 21. $2+5+10=17$

22. $5+4+8=17$ 23. $2+2+9=13$ 24. $7+3+10=20$

25. $3+6+8=17$ 26. $5+3+9=17$ 27. $5+6+10=21$

28. $4+4+8=16$ 29. $2+9+9=20$ 30. $6+2+10=18$

Name: Lesson 3-9 Adding three numbers

1. $2+4+1=7$	2. $5+6+1=12$	3. $3+8+1=12$
4. $6+3+2=11$	5. $8+2+2=12$	6. $2+2+2=6$
7. $4+6+3=13$	8. $9+2+3=14$	9. $4+3+3=10$
10. $5+5+4=14$	11. $3+4+4=11$	12. $3+6+4=13$
13. $3+4+5=12$	14. $2+5+5=12$	15. $2+3+5=10$
16. $7+2+6=15$	17. $4+4+6=14$	18. $2+8+6=16$
19. $2+6+7=15$	20. $2+4+7=13$	21. $4+7+7=18$
22. $7+3+8=18$	23. $5+4+8=17$	24. $6+2+8=16$
25. $6+5+9=20$	26. $3+9+9=21$	27. $2+7+9=18$
28. $6+4+10=20$	29. $8+3+10=21$	30. $5+2+10=17$

Name: Lesson 3-10 Adding three numbers

1. $4+3+1=8$	2. $7+4+1=12$	3. $3+3+1=7$
4. $6+5+2=13$	5. $3+6+2=11$	6. $5+4+2=11$
7. $8+2+3=13$	8. $5+3+3=11$	9. $7+2+3=12$
10. $6+4+4=14$	11. $2+3+4=10$	12. $2+8+4=14$
13. $2+6+5=13$	14. $8+3+5=16$	15. $2+9+5=16$
16. $5+4+6=15$	17. $3+2+6=11$	18. $4+6+6=16$
19. $2+6+7=15$	20. $2+4+7=13$	21. $2+5+7=14$
22. $5+2+8=15$	23. $2+7+8=17$	24. $5+5+8=18$
25. $9+2+9=20$	26. $7+3+9=19$	27. $3+7+9=19$
28. $4+7+10=21$	29. $5+6+10=21$	30. $3+8+10=21$

Name: Lesson 4-1 Adding and subtracting three numbers

1. $5-2+1=4$	2. $7-3+2=6$	3. $10-9+3=4$
4. $9-6+1=4$	5. $6-5+2=3$	6. $10-7+3=6$
7. $7-5+1=3$	8. $4-3+2=3$	9. $5-4+3=4$
10. $9-8+1=2$	11. $10-8+2=4$	12. $8-4+3=7$
13. $6-3+1=4$	14. $10-6+2=6$	15. $9-7+3=5$
16. $8-6+1=3$	17. $9-5+2=6$	18. $7-6+3=4$
19. $10-4+1=7$	20. $8-5+2=5$	21. $10-2+3=11$
22. $4-2+1=3$	23. $9-2+2=9$	24. $6-4+3=5$
25. $8-2+1=7$	26. $10-6+2=6$	27. $5-3+3=5$
28. $9-3+1=7$	29. $6-2+2=6$	30. $8-7+3=4$

Name: Lesson 4-2 Adding and subtracting three numbers

1. $9-3+2=8$	2. $7-4+3=6$	3. $10-4+4=10$
4. $6-3+2=5$	5. $5-2+3=6$	6. $9-7+4=6$
7. $10-8+2=4$	8. $7-6+3=4$	9. $10-9+4=5$
10. $8-4+2=6$	11. $4-3+3=4$	12. $9-5+4=8$
13. $9-7+2=4$	14. $8-5+3=6$	15. $6-5+4=5$
16. $5-4+2=3$	17. $5-3+3=5$	18. $10-7+4=7$
19. $7-2+2=7$	20. $9-2+3=10$	21. $3-2+4=5$
22. $10-3+2=9$	23. $6-4+3=5$	24. $8-2+4=10$
25. $6-2+2=6$	26. $7-5+3=5$	27. $4-2+4=6$
28. $10-5+2=7$	29. $9-6+3=6$	30. $10-6+4=8$

Name:

Lesson 4-3 Adding and subtracting three numbers

1. $6-4+3=5$	2. $10-9+4=5$	3. $6-5+5=6$
4. $9-4+3=8$	5. $10-8+4=6$	6. $5-2+5=8$
7. $7-2+3=8$	8. $8-6+4=6$	9. $7-4+5=8$
10. $10-6+3=7$	11. $9-8+4=5$	12. $10-4+5=11$
13. $8-4+3=7$	14. $6-2+4=8$	15. $9-7+5=7$
16. $3-2+3=4$	17. $7-5+4=6$	18. $10-3+5=12$
19. $9-5+3=7$	20. $10-5+4=9$	21. $8-2+5=11$
22. $5-4+3=4$	23. $4-2+4=6$	24. $10-7+5=8$
25. $7-3+3=7$	26. $8-5+4=7$	27. $7-6+5=6$
28. $10-2+3=11$	29. $9-3+4=10$	30. $9-6+5=8$

Name:

Lesson 4-4 Adding and subtracting three numbers

1. $10-3+4=11$	2. $7-5+5=7$	3. $8-4+6=10$
4. $5-4+4=5$	5. $10-5+5=10$	6. $6-2+6=10$
7. $3-2+4=5$	8. $9-5+5=9$	9. $5-3+6=8$
10. $6-4+4=6$	11. $10-9+5=6$	12. $4-2+6=8$
13. $8-7+4=5$	14. $5-2+5=8$	15. $9-6+6=9$
16. $8-5+4=7$	17. $7-3+5=9$	18. $10-4+6=12$
19. $9-2+4=11$	20. $9-4+5=10$	21. $7-6+6=7$
22. $10-8+4=6$	23. $7-4+5=8$	24. $10-7+6=9$
25. $4-3+4=5$	26. $10-2+5=13$	27. $8-6+6=8$
28. $9-3+4=10$	29. $8-3+5=10$	30. $8-2+6=12$

Name:

Lesson 4-5 Adding and subtracting three numbers

1. $9-7+5=7$	2. $6-5+6=7$	3. $5-2+7=10$
4. $8-7+5=6$	5. $10-8+6=8$	6. $10-9+7=8$
7. $4-2+5=7$	8. $8-3+6=11$	9. $9-5+7=11$
10. $9-6+5=8$	11. $10-5+6=11$	12. $5-3+7=9$
13. $8-5+5=8$	14. $7-2+6=11$	15. $10-7+7=10$
16. $7-4+5=8$	17. $10-4+6=12$	18. $8-4+7=11$
19. $6-3+5=8$	20. $7-5+6=8$	21. $6-2+7=11$
22. $10-3+5=12$	23. $4-3+6=7$	24. $10-6+7=11$
25. $9-2+5=12$	26. $9-3+6=12$	27. $8-2+7=13$
28. $8-6+5=7$	29. $6-4+6=8$	30. $9-4+7=12$

Name:

Lesson 4-6 Adding and subtracting three numbers

1. $9-4+6=11$	2. $5-3+7=9$	3. $9-8+8=9$
4. $10-5+6=11$	5. $6-2+7=11$	6. $8-6+8=10$
7. $3-2+6=7$	8. $8-4+7=11$	9. $10-4+8=14$
10. $6-4+6=8$	11. $6-5+7=8$	12. $4-2+8=14$
13. $8-7+6=7$	14. $5-2+7=10$	15. $7-6+8=9$
16. $9-7+6=8$	17. $7-5+7=9$	18. $10-6+8=12$
19. $10-7+6=9$	20. $9-5+7=11$	21. $5-4+8=9$
22. $8-2+6=12$	23. $7-3+7=11$	24. $9-6+8=11$
25. $7-3+6=10$	26. $8-5+7=10$	27. $9-2+8=15$
28. $10-9+6=7$	29. $9-3+7=13$	30. $7-4+8=11$

Name: **Lesson 4-7 Adding and subtracting three numbers**

1. $5 - 4 + 7 = 8$
2. $7 - 4 + 8 = 11$
3. $8 - 4 + 9 = 13$
4. $8 - 2 + 7 = 13$
5. $10 - 7 + 8 = 11$
6. $6 - 3 + 9 = 15$
7. $10 - 8 + 7 = 9$
8. $9 - 8 + 8 = 9$
9. $5 - 2 + 9 = 12$
10. $6 - 5 + 7 = 8$
11. $9 - 7 + 8 = 10$
12. $7 - 5 + 9 = 11$
13. $8 - 3 + 7 = 12$
14. $10 - 2 + 8 = 16$
15. $5 - 3 + 9 = 11$
16. $7 - 3 + 7 = 11$
17. $4 - 3 + 8 = 9$
18. $10 - 4 + 9 = 15$
19. $6 - 2 + 7 = 11$
20. $8 - 6 + 8 = 10$
21. $10 - 9 + 9 = 10$
22. $10 - 5 + 7 = 12$
23. $9 - 4 + 8 = 13$
24. $7 - 2 + 9 = 14$
25. $8 - 7 + 7 = 8$
26. $10 - 6 + 8 = 12$
27. $6 - 4 + 9 = 11$
28. $9 - 3 + 7 = 13$
29. $9 - 2 + 8 = 15$
30. $7 - 6 + 9 = 10$

Name: **Lesson 4-8 Adding and subtracting three numbers**

1. $7 - 4 + 8 = 11$
2. $6 - 5 + 9 = 10$
3. $8 - 4 + 10 = 14$
4. $9 - 6 + 8 = 11$
5. $3 - 2 + 9 = 10$
6. $10 - 2 + 10 = 18$
7. $8 - 2 + 8 = 14$
8. $10 - 7 + 9 = 12$
9. $4 - 3 + 10 = 11$
10. $7 - 6 + 8 = 9$
11. $6 - 3 + 9 = 12$
12. $9 - 8 + 10 = 11$
13. $10 - 4 + 8 = 14$
14. $5 - 3 + 9 = 11$
15. $8 - 7 + 10 = 11$
16. $8 - 6 + 8 = 10$
17. $7 - 5 + 9 = 11$
18. $10 - 5 + 10 = 15$
19. $10 - 6 + 8 = 12$
20. $8 - 3 + 9 = 15$
21. $9 - 4 + 10 = 15$
22. $7 - 2 + 8 = 13$
23. $4 - 2 + 9 = 11$
24. $7 - 3 + 10 = 14$
25. $9 - 3 + 8 = 14$
26. $10 - 3 + 9 = 16$
27. $6 - 4 + 10 = 12$
28. $5 - 2 + 8 = 11$
29. $9 - 5 + 9 = 13$
30. $10 - 8 + 10 = 12$

Name: **Lesson 4-9 Adding and subtracting three numbers**

1. $8 - 6 + 1 = 3$
2. $6 - 4 + 1 = 3$
3. $10 - 3 + 1 = 8$
4. $4 - 2 + 2 = 4$
5. $10 - 6 + 2 = 6$
6. $9 - 8 + 2 = 3$
7. $10 - 7 + 3 = 6$
8. $9 - 7 + 3 = 5$
9. $3 - 2 + 3 = 4$
10. $5 - 4 + 4 = 5$
11. $8 - 2 + 4 = 10$
12. $7 - 4 + 4 = 7$
13. $9 - 4 + 5 = 10$
14. $7 - 6 + 5 = 6$
15. $5 - 3 + 5 = 7$
16. $5 - 2 + 6 = 9$
17. $9 - 5 + 6 = 10$
18. $6 - 2 + 6 = 10$
19. $10 - 5 + 7 = 12$
20. $7 - 5 + 7 = 9$
21. $8 - 4 + 7 = 11$
22. $10 - 2 + 8 = 16$
23. $6 - 3 + 8 = 11$
24. $10 - 9 + 8 = 9$
25. $9 - 3 + 9 = 15$
26. $8 - 5 + 9 = 12$
27. $10 - 3 + 9 = 16$
28. $4 - 1 + 10 = 13$
29. $9 - 6 + 10 = 13$
30. $8 - 6 + 10 = 12$

Name: **Lesson 4-10 Adding and subtracting three numbers**

1. $7 - 5 + 1 = 3$
2. $10 - 2 + 1 = 9$
3. $10 - 3 + 1 = 8$
4. $10 - 6 + 2 = 6$
5. $8 - 4 + 2 = 6$
6. $9 - 8 + 2 = 3$
7. $9 - 8 + 3 = 4$
8. $6 - 5 + 3 = 4$
9. $3 - 2 + 3 = 4$
10. $8 - 6 + 4 = 6$
11. $9 - 7 + 4 = 6$
12. $7 - 4 + 4 = 7$
13. $4 - 2 + 5 = 7$
14. $10 - 4 + 5 = 11$
15. $5 - 3 + 5 = 7$
16. $6 - 4 + 6 = 8$
17. $7 - 6 + 6 = 7$
18. $6 - 2 + 6 = 10$
19. $8 - 3 + 7 = 12$
20. $5 - 3 + 7 = 9$
21. $8 - 4 + 7 = 11$
22. $5 - 4 + 8 = 9$
23. $3 - 2 + 8 = 9$
24. $10 - 9 + 8 = 9$
25. $9 - 6 + 9 = 12$
26. $10 - 9 + 9 = 10$
27. $10 - 3 + 9 = 16$
28. $8 - 2 + 10 = 16$
29. $7 - 2 + 10 = 15$
30. $8 - 6 + 10 = 12$

Name: **Lesson 5-1 Adding one and two digit numbers**

1. $11 + 9 =$	2. $56 + 2 = 58$	3. $35 + 4 = 39$	4. $23 + 5 = 28$	5. $41 + 8 = 49$
6. $\overset{1}{28} + 2 = 30$	7. $\overset{1}{15} + 7 = 22$	8. $\overset{1}{67} + 3 = 70$	9. $\overset{1}{33} + 8 = 41$	10. $\overset{1}{27} + 5 = 32$
11. $\overset{1}{39} + 8 = 47$	12. $\overset{1}{54} + 8 = 62$	13. $\overset{1}{72} + 9 = 81$	14. $\overset{1}{16} + 5 = 21$	15. $\overset{1}{47} + 7 = 54$
16. $43 + 6 = 49$	17. $\overset{1}{18} + 9 = 27$	18. $\overset{1}{36} + 7 = 43$	19. $\overset{1}{22} + 9 = 31$	20. $\overset{1}{44} + 9 = 53$
21. $\overset{1}{54} + 9 = 63$	22. $\overset{1}{48} + 3 = 51$	23. $\overset{1}{66} + 8 = 74$	24. $\overset{1}{34} + 7 = 41$	25. $\overset{1}{59} + 5 = 64$
26. $\overset{1}{68} + 7 = 75$	27. $\overset{1}{67} + 5 = 72$	28. $\overset{1}{47} + 5 = 52$	29. $\overset{1}{56} + 6 = 62$	30. $\overset{1}{29} + 9 = 38$

Name: **Lesson 5-2 Adding one and two digit numbers**

1. $25 + 3 = 28$	2. $52 + 6 = 58$	3. $\overset{1}{18} + 2 = 20$	4. $37 + 2 = 39$	5. $\overset{1}{46} + 7 = 53$
6. $\overset{1}{68} + 5 = 73$	7. $\overset{1}{17} + 4 = 21$	8. $\overset{1}{78} + 5 = 83$	9. $\overset{1}{25} + 9 = 34$	10. $\overset{1}{58} + 4 = 62$
11. $\overset{1}{37} + 6 = 43$	12. $\overset{1}{12} + 9 = 21$	13. $\overset{1}{48} + 3 = 51$	14. $\overset{1}{64} + 6 = 70$	15. $\overset{1}{17} + 3 = 20$
16. $\overset{1}{76} + 4 = 80$	17. $\overset{1}{36} + 8 = 44$	18. $\overset{1}{35} + 8 = 43$	19. $\overset{1}{13} + 8 = 21$	20. $\overset{1}{28} + 5 = 33$
21. $\overset{1}{65} + 6 = 71$	22. $\overset{1}{67} + 4 = 71$	23. $\overset{1}{65} + 6 = 71$	24. $\overset{1}{45} + 6 = 51$	25. $\overset{1}{37} + 7 = 44$
26. $\overset{1}{27} + 9 = 36$	27. $\overset{1}{38} + 5 = 43$	28. $\overset{1}{59} + 9 = 68$	29. $\overset{1}{74} + 7 = 81$	30. $\overset{1}{15} + 9 = 24$

Name: **Lesson 5-3 Adding one and two digit numbers**

1. $14 + 4 = 18$	2. $65 + 3 = 68$	3. $11 + 7 = 18$	4. $46 + 2 = 48$	5. $71 + 8 = 79$
6. $\overset{1}{36} + 6 = 42$	7. $\overset{1}{78} + 8 = 86$	8. $\overset{1}{32} + 8 = 40$	9. $\overset{1}{18} + 3 = 21$	10. $\overset{1}{28} + 3 = 31$
11. $\overset{1}{37} + 5 = 42$	12. $\overset{1}{76} + 8 = 84$	13. $\overset{1}{48} + 8 = 56$	14. $\overset{1}{53} + 7 = 60$	15. $\overset{1}{16} + 4 = 20$
16. $\overset{1}{64} + 8 = 72$	17. $\overset{1}{15} + 9 = 24$	18. $\overset{1}{15} + 6 = 21$	19. $\overset{1}{27} + 6 = 33$	20. $\overset{1}{44} + 8 = 52$
21. $\overset{1}{72} + 9 = 81$	22. $\overset{1}{58} + 4 = 62$	23. $\overset{1}{76} + 7 = 83$	24. $\overset{1}{38} + 7 = 45$	25. $\overset{1}{58} + 5 = 63$
26. $\overset{1}{28} + 7 = 35$	27. $\overset{1}{69} + 7 = 76$	28. $\overset{1}{38} + 6 = 44$	29. $\overset{1}{29} + 8 = 37$	30. $\overset{1}{28} + 2 = 30$

Name: **Lesson 5-4 Adding one and two digit numbers**

1. $17 + 1 = 18$	2. $30 + 8 = 38$	3. $76 + 2 = 78$	4. $23 + 5 = 28$	5. $\overset{1}{58} + 2 = 60$
6. $\overset{1}{13} + 7 = 20$	7. $\overset{1}{68} + 8 = 76$	8. $\overset{1}{45} + 8 = 53$	9. $\overset{1}{18} + 7 = 25$	10. $\overset{1}{36} + 4 = 40$
11. $\overset{1}{78} + 4 = 82$	12. $\overset{1}{28} + 4 = 32$	13. $\overset{1}{15} + 5 = 20$	14. $\overset{1}{64} + 7 = 71$	15. $\overset{1}{47} + 7 = 54$
16. $\overset{1}{55} + 6 = 61$	17. $\overset{1}{16} + 5 = 21$	18. $\overset{1}{37} + 4 = 41$	19. $\overset{1}{76} + 8 = 84$	20. $\overset{1}{12} + 9 = 21$
21. $\overset{1}{48} + 8 = 56$	22. $\overset{1}{67} + 6 = 73$	23. $\overset{1}{54} + 6 = 60$	24. $\overset{1}{26} + 5 = 31$	25. $\overset{1}{58} + 5 = 63$
26. $\overset{1}{25} + 5 = 30$	27. $\overset{1}{56} + 9 = 65$	28. $\overset{1}{43} + 8 = 51$	29. $\overset{1}{36} + 8 = 44$	30. $\overset{1}{17} + 6 = 23$

Name: **Lesson 5-5 Adding one and two digit numbers**

1. 26 + 2 = 28	2. 64 + 4 = 68	3. 71 + 7 = 78	4. 37 + 5 = 42	5. 18 + 0 = 18
6. 47 + 7 = 54	7. 13 + 9 = 22	8. 56 + 4 = 60	9. 64 + 7 = 71	10. 13 + 7 = 20
11. 78 + 3 = 81	12. 18 + 7 = 25	13. 48 + 3 = 51	14. 56 + 8 = 64	15. 24 + 8 = 32
16. 38 + 2 = 40	17. 27 + 8 = 35	18. 67 + 6 = 73	19. 45 + 5 = 50	20. 18 + 5 = 23
21. 54 + 8 = 62	22. 75 + 6 = 81	23. 38 + 8 = 46	24. 62 + 8 = 70	25. 17 + 8 = 25
26. 15 + 8 = 23	27. 37 + 5 = 42	28. 58 + 4 = 62	29. 24 + 6 = 30	30. 44 + 9 = 53

Name: **Lesson 5-6 Adding one and two digit numbers**

1. 43 + 5 = 48	2. 11 + 8 = 19	3. 60 + 8 = 68	4. 34 + 4 = 38	5. 57 + 2 = 59
6. 78 + 2 = 80	7. 14 + 9 = 23	8. 23 + 7 = 30	9. 16 + 5 = 21	10. 38 + 6 = 44
11. 25 + 6 = 31	12. 41 + 9 = 50	13. 67 + 5 = 72	14. 58 + 5 = 63	15. 17 + 4 = 21
16. 76 + 8 = 84	17. 15 + 6 = 21	18. 38 + 7 = 45	19. 14 + 6 = 20	20. 52 + 8 = 60
21. 43 + 8 = 51	22. 67 + 6 = 73	23. 26 + 7 = 33	24. 74 + 7 = 81	25. 18 + 3 = 21
26. 32 + 9 = 41	27. 15 + 8 = 23	28. 75 + 5 = 80	29. 46 + 8 = 54	30. 67 + 7 = 74

Name: **Lesson 5-7 Adding one and two digit numbers**

1. 65 + 3 = 68	2. 42 + 6 = 48	3. 18 + 1 = 19	4. 37 + 2 = 39	5. 16 + 3 = 19
6. 53 + 7 = 60	7. 72 + 9 = 81	8. 25 + 7 = 32	9. 14 + 6 = 20	10. 38 + 4 = 42
11. 57 + 8 = 65	12. 16 + 8 = 24	13. 46 + 5 = 51	14. 25 + 6 = 31	15. 63 + 8 = 71
16. 78 + 8 = 86	17. 17 + 3 = 20	18. 35 + 8 = 43	19. 17 + 5 = 22	20. 41 + 9 = 50
21. 68 + 3 = 71	22. 77 + 6 = 83	23. 26 + 6 = 32	24. 56 + 7 = 63	25. 15 + 5 = 20
26. 27 + 4 = 31	27. 36 + 4 = 40	28. 65 + 6 = 71	29. 14 + 9 = 23	30. 78 + 6 = 84

Name: **Lesson 5-8 Adding one and two digit numbers**

1. 42 + 6 = 48	2. 64 + 4 = 68	3. 73 + 5 = 78	4. 27 + 1 = 28	5. 56 + 4 = 60
6. 38 + 3 = 41	7. 15 + 9 = 24	8. 17 + 4 = 21	9. 32 + 9 = 41	10. 17 + 8 = 25
11. 43 + 7 = 50	12. 28 + 6 = 34	13. 15 + 7 = 22	14. 54 + 8 = 62	15. 68 + 5 = 73
16. 74 + 6 = 80	17. 18 + 8 = 26	18. 38 + 2 = 40	19. 61 + 9 = 70	20. 24 + 7 = 31
21. 57 + 6 = 63	22. 45 + 5 = 50	23. 73 + 7 = 80	24. 16 + 6 = 22	25. 17 + 5 = 22
26. 43 + 8 = 51	27. 13 + 8 = 21	28. 57 + 8 = 65	29. 26 + 7 = 33	30. 36 + 6 = 42

Name: **Lesson 5-9 Adding one and two digit numbers**

1. $48 + 1 = 49$	2. $\overset{1}{17} + 4 = 21$	3. $62 + 6 = 68$	4. $14 + 4 = 18$	5. $\overset{1}{32} + 8 = 40$
6. $21 + 8 = 29$	7. $\overset{1}{76} + 6 = 82$	8. $\overset{1}{54} + 8 = 62$	9. $\overset{1}{28} + 3 = 31$	10. $\overset{1}{36} + 5 = 41$
11. $\overset{1}{12} + 9 = 21$	12. $\overset{1}{18} + 7 = 25$	13. $\overset{1}{65} + 6 = 71$	14. $\overset{1}{45} + 7 = 52$	15. $\overset{1}{53} + 8 = 61$
16. $\overset{1}{73} + 7 = 80$	17. $\overset{1}{15} + 5 = 20$	18. $\overset{1}{34} + 7 = 41$	19. $\overset{1}{18} + 6 = 24$	20. $\overset{1}{67} + 9 = 76$
21. $\overset{1}{23} + 8 = 31$	22. $\overset{1}{74} + 7 = 81$	23. $\overset{1}{46} + 6 = 52$	24. $\overset{1}{58} + 6 = 64$	25. $\overset{1}{77} + 8 = 85$
26. $\overset{1}{58} + 4 = 62$	27. $\overset{1}{25} + 9 = 34$	28. $\overset{1}{17} + 4 = 21$	29. $\overset{1}{18} + 4 = 22$	30. $\overset{1}{33} + 9 = 42$

Name: **Lesson 5-10 Adding one and two digit numbers**

1. $37 + 2 = 39$	2. $\overset{1}{28} + 2 = 30$	3. $28 + 1 = 29$	4. $75 + 4 = 79$	5. $56 + 3 = 59$
6. $\overset{1}{63} + 8 = 71$	7. $\overset{1}{27} + 4 = 31$	8. $\overset{1}{17} + 4 = 21$	9. $\overset{1}{17} + 8 = 25$	10. $\overset{1}{13} + 9 = 22$
11. $\overset{1}{48} + 2 = 50$	12. $\overset{1}{28} + 3 = 31$	13. $\overset{1}{28} + 4 = 32$	14. $\overset{1}{36} + 6 = 42$	15. $\overset{1}{65} + 9 = 74$
16. $\overset{1}{78} + 3 = 81$	17. $\overset{1}{15} + 8 = 23$	18. $\overset{1}{16} + 9 = 25$	19. $\overset{1}{48} + 9 = 57$	20. $\overset{1}{24} + 6 = 30$
21. $\overset{1}{53} + 8 = 61$	22. $\overset{1}{78} + 4 = 82$	23. $\overset{1}{78} + 5 = 83$	24. $\overset{1}{67} + 4 = 71$	25. $\overset{1}{17} + 9 = 26$
26. $\overset{1}{26} + 6 = 32$	27. $\overset{1}{43} + 9 = 52$	28. $\overset{1}{43} + 9 = 52$	29. $\overset{1}{58} + 5 = 63$	30. $\overset{1}{32} + 8 = 40$

Name: **Lesson 6-1 Subtracting one and two digit numbers**

1. $60 - 8 = 52$	2. $21 - 6 = 15$	3. $71 - 4 = 67$	4. $45 - 9 = 36$	5. $50 - 7 = 43$
6. $39 - 2 = 37$	7. $81 - 5 = 76$	8. $13 - 5 = 8$	9. $82 - 8 = 74$	10. $42 - 6 = 36$
11. $63 - 8 = 55$	12. $51 - 7 = 44$	13. $70 - 7 = 63$	14. $31 - 9 = 22$	15. $26 - 8 = 18$
16. $12 - 9 = 3$	17. $70 - 3 = 67$	18. $87 - 9 = 78$	19. $42 - 5 = 37$	20. $62 - 4 = 58$
21. $36 - 9 = 27$	22. $11 - 8 = 3$	23. $20 - 5 = 15$	24. $50 - 8 = 42$	25. $82 - 7 = 75$
26. $65 - 8 = 57$	27. $53 - 7 = 46$	28. $43 - 9 = 34$	29. $73 - 6 = 67$	30. $31 - 8 = 23$

Name: **Lesson 6-2 Subtracting one and two digit numbers**

1. $11 - 7 = 4$	2. $31 - 8 = 23$	3. $61 - 6 = 55$	4. $27 - 9 = 18$	5. $84 - 6 = 78$
6. $71 - 9 = 62$	7. $42 - 4 = 38$	8. $51 - 3 = 48$	9. $12 - 5 = 7$	10. $21 - 9 = 12$
11. $32 - 6 = 26$	12. $72 - 7 = 65$	13. $60 - 5 = 55$	14. $81 - 8 = 73$	15. $40 - 4 = 36$
16. $72 - 8 = 64$	17. $20 - 9 = 11$	18. $51 - 7 = 44$	19. $63 - 6 = 57$	20. $26 - 8 = 18$
21. $45 - 7 = 38$	22. $32 - 9 = 23$	23. $71 - 3 = 68$	24. $81 - 5 = 76$	25. $11 - 4 = 7$
26. $76 - 9 = 67$	27. $24 - 6 = 18$	28. $83 - 8 = 75$	29. $50 - 3 = 47$	30. $60 - 6 = 54$

Name:

Lesson 6-3 Subtracting one and two digit numbers

1. 74 − 9 = 65	2. 42 − 8 = 34	3. 63 − 6 = 57	4. 51 − 7 = 44	5. 12 − 9 = 3
6. 23 − 8 = 15	7. 82 − 7 = 75	8. 31 − 4 = 27	9. 35 − 9 = 26	10. 41 − 3 = 38
11. 51 − 2 = 49	12. 21 − 9 = 12	13. 14 − 6 = 8	14. 73 − 7 = 66	15. 60 − 5 = 55
16. 82 − 6 = 76	17. 14 − 7 = 7	18. 34 − 8 = 26	19. 50 − 4 = 46	20. 26 − 9 = 17
21. 41 − 8 = 33	22. 60 − 7 = 53	23. 72 − 4 = 68	24. 81 − 9 = 72	25. 51 − 6 = 45
26. 11 − 4 = 7	27. 31 − 5 = 26	28. 65 − 7 = 58	29. 20 − 5 = 15	30. 75 − 8 = 67

Name:

Lesson 6-4 Subtracting one and two digit numbers

1. 54 − 6 = 48	2. 30 − 3 = 27	3. 43 − 7 = 36	4. 72 − 9 = 63	5. 31 − 7 = 24
6. 31 − 8 = 23	7. 61 − 9 = 52	8. 81 − 5 = 76	9. 53 − 4 = 49	10. 64 − 6 = 58
11. 16 − 7 = 9	12. 90 − 6 = 84	13. 21 − 9 = 12	14. 82 − 8 = 74	15. 31 − 5 = 26
16. 35 − 7 = 28	17. 23 − 6 = 17	18. 44 − 9 = 35	19. 33 − 5 = 28	20. 50 − 8 = 42
21. 30 − 9 = 21	22. 71 − 7 = 64	23. 63 − 8 = 55	24. 80 − 6 = 74	25. 61 − 7 = 54
26. 54 − 9 = 45	27. 22 − 4 = 18	28. 31 − 4 = 27	29. 27 − 9 = 18	30. 81 − 8 = 73

Name:

Lesson 6-5 Subtracting one and two digit numbers

1. 13 − 9 = 4	2. 51 − 8 = 43	3. 21 − 3 = 18	4. 84 − 9 = 75	5. 72 − 8 = 64
6. 30 − 8 = 22	7. 70 − 7 = 63	8. 42 − 4 = 38	9. 21 − 3 = 18	10. 71 − 8 = 63
11. 41 − 9 = 32	12. 24 − 6 = 18	13. 61 − 5 = 56	14. 50 − 7 = 43	15. 30 − 6 = 24
16. 82 − 7 = 75	17. 32 − 8 = 24	18. 21 − 3 = 18	19. 42 − 9 = 33	20. 21 − 5 = 16
21. 52 − 5 = 47	22. 80 − 4 = 76	23. 71 − 6 = 65	24. 61 − 4 = 57	25. 15 − 7 = 8
26. 21 − 6 = 15	27. 83 − 7 = 76	28. 43 − 8 = 35	29. 56 − 8 = 48	30. 63 − 6 = 57

Name:

Lesson 6-6 Subtracting one and two digit numbers

1. 63 − 8 = 55	2. 43 − 9 = 34	3. 72 − 7 = 65	4. 21 − 5 = 16	5. 32 − 6 = 26
6. 22 − 5 = 17	7. 80 − 6 = 74	8. 53 − 8 = 45	9. 41 − 4 = 37	10. 71 − 5 = 66
11. 14 − 9 = 5	12. 39 − 5 = 34	13. 83 − 7 = 76	14. 20 − 2 = 18	15. 61 − 4 = 57
16. 31 − 9 = 22	17. 14 − 5 = 9	18. 60 − 8 = 52	19. 31 − 3 = 28	20. 42 − 6 = 36
21. 54 − 8 = 46	22. 74 − 7 = 67	23. 21 − 3 = 18	24. 80 − 9 = 71	25. 51 − 7 = 44
26. 40 − 3 = 37	27. 23 − 6 = 17	28. 65 − 9 = 56	29. 37 − 8 = 29	30. 33 − 9 = 24

Name:

Lesson 6-7 Subtracting one and two digit numbers

1. $24 - 7 = 17$	2. $83 - 8 = 75$	3. $44 - 9 = 35$	4. $72 - 7 = 65$	5. $65 - 6 = 59$
6. $32 - 8 = 24$	7. $12 - 5 = 7$	8. $52 - 6 = 46$	9. $52 - 4 = 48$	10. $73 - 5 = 68$
11. $15 - 9 = 6$	12. $31 - 3 = 28$	13. $23 - 7 = 16$	14. $64 - 8 = 56$	15. $50 - 3 = 47$
16. $80 - 8 = 72$	17. $11 - 2 = 9$	18. $43 - 9 = 34$	19. $26 - 9 = 17$	20. $31 - 8 = 23$
21. $51 - 9 = 42$	22. $71 - 7 = 64$	23. $81 - 8 = 73$	24. $61 - 5 = 56$	25. $32 - 6 = 26$
26. $30 - 6 = 24$	27. $52 - 3 = 49$	28. $86 - 9 = 77$	29. $40 - 9 = 31$	30. $83 - 7 = 76$

Name:

Lesson 6-8 Subtracting one and two digit numbers

1. $51 - 6 = 45$	2. $16 - 9 = 7$	3. $16 - 9 = 7$	4. $34 - 8 = 26$	5. $72 - 7 = 65$
6. $64 - 9 = 55$	7. $40 - 2 = 38$	8. $30 - 7 = 23$	9. $85 - 9 = 76$	10. $51 - 8 = 43$
11. $21 - 4 = 17$	12. $66 - 8 = 58$	13. $66 - 9 = 57$	14. $16 - 7 = 9$	15. $31 - 6 = 25$
16. $87 - 9 = 78$	17. $35 - 7 = 28$	18. $35 - 7 = 28$	19. $53 - 5 = 48$	20. $23 - 5 = 18$
21. $40 - 9 = 31$	22. $22 - 3 = 19$	23. $23 - 4 = 19$	24. $71 - 8 = 63$	25. $81 - 4 = 77$
26. $20 - 4 = 16$	27. $12 - 8 = 4$	28. $21 - 5 = 16$	29. $43 - 5 = 38$	30. $62 - 6 = 56$

Name:

Lesson 6-9 Subtracting one and two digit numbers

1. $15 - 6 = 9$	2. $34 - 9 = 25$	3. $62 - 5 = 57$	4. $61 - 6 = 55$	5. $94 - 8 = 86$
6. $43 - 8 = 35$	7. $71 - 5 = 66$	8. $23 - 8 = 15$	9. $32 - 5 = 27$	10. $74 - 5 = 69$
11. $70 - 5 = 65$	12. $21 - 7 = 14$	13. $64 - 4 = 60$	14. $91 - 9 = 82$	15. $42 - 8 = 34$
16. $41 - 9 = 32$	17. $80 - 5 = 75$	18. $82 - 7 = 75$	19. $53 - 5 = 48$	20. $35 - 5 = 30$
21. $34 - 6 = 28$	22. $34 - 5 = 29$	23. $24 - 6 = 18$	24. $90 - 3 = 87$	25. $51 - 3 = 48$
26. $36 - 9 = 27$	27. $65 - 8 = 57$	28. $41 - 8 = 33$	29. $65 - 8 = 57$	30. $93 - 5 = 88$

Name:

Lesson 6-10 Subtracting one and two digit numbers

1. $75 - 8 = 67$	2. $41 - 4 = 37$	3. $63 - 9 = 54$	4. $91 - 9 = 82$	5. $86 - 8 = 78$
6. $37 - 9 = 28$	7. $23 - 5 = 18$	8. $51 - 8 = 43$	9. $70 - 5 = 65$	10. $61 - 7 = 54$
11. $23 - 7 = 16$	12. $46 - 7 = 39$	13. $92 - 5 = 87$	14. $51 - 7 = 44$	15. $82 - 5 = 77$
16. $32 - 6 = 26$	17. $57 - 9 = 48$	18. $87 - 9 = 78$	19. $32 - 8 = 24$	20. $43 - 7 = 36$
21. $60 - 8 = 52$	22. $90 - 4 = 86$	23. $24 - 7 = 17$	24. $74 - 5 = 69$	25. $61 - 2 = 59$
26. $73 - 5 = 68$	27. $42 - 7 = 35$	28. $53 - 6 = 47$	29. $70 - 3 = 67$	30. $36 - 9 = 27$

Name: **Lesson 7-1** **Adding three numbers**

1. $14+7+1=22$	2. $23+9+2=34$	3. $16+9+3=28$
4. $18+4+1=23$	5. $15+8+2=25$	6. $29+9+3=41$
7. $27+5+1=33$	8. $16+7+2=25$	9. $19+7+3=29$
10. $16+6+1=23$	11. $29+5+2=36$	12. $18+8+3=29$
13. $25+9+1=35$	14. $18+6+2=26$	15. $27+8+3=38$
16. $26+8+1=35$	17. $27+7+2=36$	18. $14+6+3=23$
19. $22+7+1=30$	20. $15+9+2=36$	21. $24+7+3=34$
22. $19+4+1=24$	23. $24+8+2=34$	24. $15+6+3=24$
25. $27+5+1=33$	26. $19+6+2=27$	27. $28+3+3=34$
28. $26+8+1=35$	29. $29+3+2=34$	30. $15+7+3=25$

Name: **Lesson 7-2** **Adding three numbers**

1. $28+5+2=35$	2. $27+3+3=33$	3. $22+8+4=34$
4. $14+5+2=21$	5. $18+8+3=29$	6. $18+3+4=25$
7. $18+7+2=27$	8. $19+5+3=27$	9. $15+7+4=26$
10. $29+3+2=34$	11. $24+7+3=34$	12. $27+4+4=35$
13. $19+9+2=30$	14. $19+3+3=25$	15. $16+6+4=26$
16. $15+4+2=21$	17. $28+7+3=38$	18. $25+6+4=35$
19. $17+7+2=26$	20. $15+5+3=23$	21. $19+7+4=30$
22. $13+7+2=22$	23. $26+7+3=36$	24. $13+8+4=25$
25. $29+2+2=33$	26. $19+8+3=30$	27. $15+7+4=26$
28. $16+5+2=23$	29. $27+5+3=35$	30. $29+2+4=35$

Name: **Lesson 7-3** **Adding three numbers**

1. $27+7+3=37$	2. $26+7+4=37$	3. $38+8+5=51$
4. $38+4+3=45$	5. $35+6+4=45$	6. $29+3+5=37$
7. $29+3+3=35$	8. $18+6+4=28$	9. $26+9+5=40$
10. $16+5+3=24$	11. $15+7+4=26$	12. $19+9+5=33$
13. $39+7+3=49$	14. $27+8+4=39$	15. $23+8+5=36$
16. $17+5+3=25$	17. $32+7+4=43$	18. $17+6+5=28$
19. $26+4+3=33$	20. $17+7+4=28$	21. $33+8+5=46$
22. $19+7+3=29$	23. $28+5+4=37$	24. $13+9+5=27$
25. $33+5+3=41$	26. $37+4+4=45$	27. $24+8+5=37$
28. $17+9+3=29$	29. $19+3+4=26$	30. $35+5+5=45$

Name: **Lesson 7-4** **Adding three numbers**

1. $18+3+4=25$	2. $29+8+5=42$	3. $26+9+6=41$
4. $29+5+4=38$	5. $27+8+5=40$	6. $15+7+6=28$
7. $17+4+4=25$	8. $38+3+5=46$	9. $25+8+6=39$
10. $35+6+4=45$	11. $34+5+5=44$	12. $32+8+6=46$
13. $16+8+4=28$	14. $18+7+5=30$	15. $28+4+6=38$
16. $19+9+4=32$	17. $21+3+5=29$	18. $33+5+6=44$
19. $26+8+4=38$	20. $27+6+5=38$	21. $17+7+6=30$
22. $17+9+4=30$	23. $37+7+5=49$	24. $29+2+6=37$
25. $24+5+4=33$	26. $19+7+5=31$	27. $18+6+6=30$
28. $38+3+4=45$	29. $22+6+5=33$	30. $34+8+6=48$

Name: Lesson 7-5 Adding three numbers

1. $28+7+5=40$	2. $33+9+6=48$	3. $26+8+7=41$
4. $34+9+5=48$	5. $45+9+6=60$	6. $39+9+7=45$
7. $38+8+5=51$	8. $29+6+6=41$	9. $25+6+7=38$
10. $29+5+5=39$	11. $28+9+6=43$	12. $47+6+7=60$
13. $42+9+5=56$	14. $35+8+6=49$	15. $36+7+7=50$
16. $25+7+5=37$	17. $28+3+6=37$	18. $27+3+7=37$
19. $38+4+5=47$	20. $39+2+6=47$	21. $39+4+7=50$
22. $37+6+5=48$	23. $44+5+6=55$	24. $48+3+7=58$
25. $46+6+5=57$	26. $24+8+6=38$	27. $27+5+7=39$
28. $27+4+5=36$	29. $36+7+6=49$	30. $48+6+7=61$

Name: Lesson 7-6 Adding three numbers

1. $46+8+6=60$	2. $25+7+7=39$	3. $48+3+8=59$
4. $39+3+6=48$	5. $39+7+7=53$	6. $37+6+8=51$
7. $36+5+6=47$	8. $29+4+7=40$	9. $34+7+8=49$
10. $44+8+6=58$	11. $42+9+7=58$	12. $28+7+8=43$
13. $17+6+6=29$	14. $46+6+7=59$	15. $23+9+8=40$
16. $18+9+6=33$	17. $35+9+7=51$	18. $47+4+8=59$
19. $37+8+6=51$	20. $28+9+7=44$	21. $39+6+8=53$
22. $46+9+6=61$	23. $48+5+7=60$	24. $46+7+8=61$
25. $23+8+6=37$	26. $35+6+7=48$	27. $39+5+8=52$
28. $29+3+6=38$	29. $27+9+7=43$	30. $47+5+8=60$

Name: Lesson 7-7 Adding three numbers

1. $38+9+7=54$	2. $49+7+8=62$	3. $42+9+9=60$
4. $49+3+7=59$	5. $55+7+8=70$	6. $58+4+9=71$
7. $48+3+7=58$	8. $46+9+8=63$	9. $49+4+9=62$
10. $59+5+7=71$	11. $38+5+8=51$	12. $57+7+9=73$
13. $36+8+7=51$	14. $53+9+8=70$	15. $35+9+9=53$
16. $49+9+7=65$	17. $46+7+8=61$	18. $49+2+9=60$
19. $55+6+7=68$	20. $57+9+8=74$	21. $59+6+9=74$
22. $44+7+7=58$	23. $34+8+8=50$	24. $47+8+9=64$
25. $58+7+7=72$	26. $46+6+8=60$	27. $37+9+9=55$
28. $33+8+7=48$	29. $57+5+8=70$	30. $36+8+9=53$

Name: Lesson 7-8 Adding three numbers

1. $56+9+8=73$	2. $37+8+9=54$	3. $47+6+10=63$
4. $48+3+8=59$	5. $48+8+9=65$	6. $59+7+10=76$
7. $39+2+8=49$	8. $45+6+9=60$	9. $36+6+10=52$
10. $57+7+8=72$	11. $58+9+9=76$	12. $45+7+10=62$
13. $39+4+8=51$	14. $49+3+9=61$	15. $55+8+10=73$
16. $44+9+8=61$	17. $36+8+9=53$	18. $47+6+10=63$
19. $35+8+8=51$	20. $59+9+9=77$	21. $38+4+10=52$
22. $45+7+8=60$	23. $36+5+9=50$	24. $48+9+10=67$
25. $58+5+8=71$	26. $33+8+9=50$	27. $39+5+10=54$
28. $31+9+8=48$	29. $46+9+9=64$	30. $44+6+10=60$

Name: **Lesson 7-9 Adding three numbers**

1. $35 + 9 + 1 = 42$
2. $38 + 5 + 1 = 44$
3. $39 + 3 + 1 = 43$
4. $48 + 8 + 2 = 58$
5. $57 + 6 + 2 = 65$
6. $36 + 8 + 2 = 46$
7. $57 + 4 + 3 = 64$
8. $45 + 6 + 3 = 54$
9. $48 + 4 + 3 = 55$
10. $49 + 4 + 4 = 57$
11. $33 + 9 + 4 = 46$
12. $57 + 8 + 4 = 69$
13. $58 + 9 + 5 = 72$
14. $49 + 5 + 5 = 59$
15. $46 + 7 + 5 = 58$
16. $36 + 6 + 6 = 48$
17. $56 + 5 + 6 = 67$
18. $36 + 7 + 6 = 49$
19. $39 + 6 + 7 = 52$
20. $48 + 3 + 7 = 58$
21. $45 + 6 + 7 = 58$
22. $44 + 7 + 8 = 59$
23. $37 + 7 + 8 = 52$
24. $59 + 2 + 8 = 69$
25. $57 + 5 + 9 = 71$
26. $44 + 9 + 9 = 62$
27. $39 + 8 + 9 = 56$
28. $49 + 9 + 10 = 68$
29. $53 + 6 + 10 = 69$
30. $49 + 7 + 10 = 66$

Name: **Lesson 7-10 Adding three numbers**

1. $54 + 7 + 1 = 62$
2. $48 + 8 + 1 = 57$
3. $39 + 7 + 1 = 47$
4. $44 + 9 + 2 = 55$
5. $58 + 3 + 2 = 63$
6. $58 + 9 + 2 = 69$
7. $46 + 8 + 3 = 57$
8. $39 + 5 + 3 = 47$
9. $49 + 9 + 3 = 61$
10. $59 + 2 + 4 = 65$
11. $59 + 4 + 4 = 67$
12. $38 + 4 + 4 = 46$
13. $57 + 4 + 5 = 66$
14. $36 + 6 + 5 = 47$
15. $59 + 4 + 5 = 68$
16. $44 + 8 + 6 = 58$
17. $45 + 6 + 6 = 57$
18. $47 + 6 + 6 = 59$
19. $36 + 9 + 7 = 52$
20. $37 + 8 + 7 = 52$
21. $33 + 8 + 7 = 48$
22. $58 + 7 + 8 = 73$
23. $53 + 9 + 8 = 70$
24. $47 + 5 + 8 = 60$
25. $47 + 7 + 9 = 63$
26. $35 + 3 + 9 = 47$
27. $59 + 6 + 9 = 74$
28. $58 + 5 + 10 = 73$
29. $46 + 5 + 10 = 61$
30. $47 + 9 + 10 = 66$

Name: **Lesson 8-1 Adding and subtracting three numbers**

1. $14 - 6 + 1 = 9$
2. $23 - 6 + 2 = 19$
3. $36 - 9 + 3 = 30$
4. $25 - 7 + 1 = 19$
5. $37 - 9 + 2 = 30$
6. $24 - 8 + 3 = 19$
7. $21 - 6 + 1 = 16$
8. $13 - 4 + 2 = 11$
9. $16 - 6 + 3 = 13$
10. $31 - 5 + 1 = 27$
11. $22 - 5 + 2 = 19$
12. $24 - 6 + 3 = 21$
13. $26 - 7 + 1 = 20$
14. $33 - 7 + 2 = 28$
15. $31 - 5 + 3 = 29$
16. $32 - 8 + 1 = 25$
17. $31 - 8 + 2 = 25$
18. $12 - 3 + 3 = 12$
19. $25 - 6 + 1 = 20$
20. $24 - 6 + 2 = 20$
21. $27 - 8 + 3 = 22$
22. $31 - 5 + 1 = 27$
23. $36 - 8 + 2 = 30$
24. $23 - 2 + 3 = 24$
25. $28 - 9 + 1 = 20$
26. $26 - 8 + 2 = 20$
27. $31 - 3 + 3 = 31$
28. $23 - 8 + 1 = 16$
29. $16 - 7 + 2 = 11$
30. $25 - 9 + 3 = 19$

Name: **Lesson 8-2 Adding and subtracting three numbers**

1. $22 - 5 + 2 = 19$
2. $25 - 7 + 3 = 21$
3. $31 - 5 + 4 = 30$
4. $11 - 4 + 2 = 9$
5. $34 - 3 + 3 = 34$
6. $17 - 8 + 4 = 13$
7. $36 - 7 + 2 = 31$
8. $22 - 4 + 3 = 21$
9. $23 - 9 + 4 = 18$
10. $24 - 6 + 2 = 20$
11. $34 - 6 + 3 = 31$
12. $36 - 8 + 4 = 32$
13. $21 - 6 + 2 = 17$
14. $23 - 8 + 3 = 18$
15. $23 - 7 + 4 = 20$
16. $14 - 9 + 2 = 7$
17. $31 - 2 + 3 = 32$
18. $15 - 6 + 4 = 13$
19. $22 - 8 + 2 = 16$
20. $17 - 9 + 3 = 11$
21. $24 - 7 + 4 = 21$
22. $16 - 8 + 2 = 10$
23. $33 - 5 + 3 = 31$
24. $21 - 9 + 4 = 16$
25. $33 - 4 + 2 = 31$
26. $11 - 8 + 3 = 6$
27. $18 - 9 + 4 = 13$
28. $32 - 9 + 2 = 25$
29. $25 - 8 + 3 = 20$
30. $31 - 7 + 4 = 28$

Name: Lesson 8-3 Adding and subtracting three numbers

1. $25-8+3=20$	2. $31-9+4=26$	3. $36-8+5=33$
4. $31-4+3=30$	5. $27-8+4=23$	6. $32-9+5=28$
7. $32-6+3=29$	8. $33-6+4=31$	9. $21-7+5=19$
10. $47-9+3=41$	11. $21-2+4=23$	12. $33-4+5=34$
13. $32-7+3=28$	14. $33-8+4=29$	15. $42-5+5=42$
16. $44-9+3=38$	17. $44-7+4=41$	18. $22-8+5=19$
19. $36-7+3=32$	20. $31-8+4=27$	21. $34-8+5=31$
22. $25-7+3=21$	23. $42-3+4=43$	24. $38-9+5=34$
25. $31-6+3=28$	26. $35-9+4=30$	27. $26-7+5=24$
28. $21-5+3=19$	29. $24-6+4=22$	30. $41-3+5=43$

Name: Lesson 8-4 Adding and subtracting three numbers

1. $24-9+4=19$	2. $21-6+5=20$	3. $45-6+6=45$
4. $37-8+4=33$	5. $36-8+5=33$	6. $33-4+6=35$
7. $24-5+4=23$	8. $32-8+5=29$	9. $24-8+6=22$
10. $35-7+4=32$	11. $43-7+5=41$	12. $35-9+6=32$
13. $43-8+4=39$	14. $32-5+5=32$	15. $23-4+6=25$
16. $41-9+4=36$	17. $28-9+5=24$	18. $31-8+6=29$
19. $32-7+4=29$	20. $23-6+5=22$	21. $46-7+6=45$
22. $41-3+4=42$	23. $42-5+5=42$	24. $32-6+6=32$
25. $24-8+4=20$	26. $35-8+5=32$	27. $41-4+6=43$
28. $44-6+4=42$	29. $31-6+5=30$	30. $22-4+6=24$

Name: Lesson 8-5 Adding and subtracting three numbers

1. $33-5+5=33$	2. $34-7+6=33$	3. $35-6+7=36$
4. $42-6+5=41$	5. $43-4+6=45$	6. $47-8+7=46$
7. $52-9+5=48$	8. $46-9+6=43$	9. $51-5+7=53$
10. $34-8+5=31$	11. $54-6+6=54$	12. $44-9+7=42$
13. $41-8+5=38$	14. $32-3+6=35$	15. $46-8+7=45$
16. $54-7+5=52$	17. $43-9+6=40$	18. $36-7+7=36$
19. $45-8+5=42$	20. $31-2+6=35$	21. $55-9+7=53$
22. $55-7+5=53$	23. $51-6+6=51$	24. $52-4+7=55$
25. $41-8+5=38$	26. $45-8+6=43$	27. $43-7+7=43$
28. $32-9+5=28$	29. $44-5+6=45$	30. $52-8+7=51$

Name: Lesson 8-6 Adding and subtracting three numbers

1. $36-9+6=33$	2. $42-7+7=42$	3. $32-3+8=37$
4. $45-7+6=44$	5. $34-8+7=33$	6. $41-7+8=42$
7. $33-8+6=31$	8. $51-5+7=53$	9. $53-4+8=57$
10. $44-6+6=44$	11. $43-5+7=45$	12. $32-7+8=33$
13. $51-4+6=53$	14. $52-4+7=55$	15. $53-6+8=55$
16. $48-9+6=45$	17. $41-3+7=45$	18. $47-9+8=46$
19. $32-8+6=30$	20. $35-7+7=35$	21. $36-8+8=36$
22. $33-7+6=32$	23. $55-8+7=54$	24. $31-2+8=37$
25. $45-9+6=42$	26. $31-9+7=29$	27. $43-9+8=42$
28. $55-7+6=54$	29. $44-5+7=46$	30. $52-5+8=55$

Name: Lesson 8-7 Adding and subtracting three numbers

1. $45-7+7=45$ 2. $65-5+8=68$ 3. $43-8+9=44$

4. $58-9+7=56$ 5. $54-6+8=56$ 6. $61-9+9=61$

7. $43-8+7=42$ 8. $56-9+8=55$ 9. $56-8+9=57$

10. $44-6+7=45$ 11. $43-5+8=46$ 12. $44-8+9=45$

13. $51-2+7=56$ 14. $61-4+8=65$ 15. $57-8+9=58$

16. $42-8+7=41$ 17. $52-8+8=52$ 18. $41-6+9=44$

19. $62-6+7=63$ 20. $61-3+8=66$ 21. $46-7+9=48$

22. $52-4+7=55$ 23. $57-9+8=56$ 24. $52-5+9=56$

25. $43-9+7=41$ 26. $42-4+8=46$ 27. $64-7+9=66$

28. $51-8+7=50$ 29. $51-5+8=54$ 30. $53-6+9=56$

Name: Lesson 8-8 Adding and subtracting three numbers

1. $43-7+8=44$ 2. $41-9+9=41$ 3. $61-9+10=62$

4. $52-5+8=55$ 5. $53-4+9=58$ 6. $52-5+10=57$

7. $65-7+8=66$ 8. $64-7+9=66$ 9. $67-8+10=69$

10. $64-5+8=67$ 11. $41-3+9=47$ 12. $42-6+10=46$

13. $51-3+8=56$ 14. $64-5+9=68$ 15. $54-7+10=57$

16. $52-9+8=51$ 17. $58-9+9=58$ 18. $44-5+10=49$

19. $42-4+8=46$ 20. $46-8+9=47$ 21. $51-5+10=56$

22. $53-8+8=53$ 23. $57-8+9=58$ 24. $62-5+10=67$

25. $67-9+8=66$ 26. $44-8+9=45$ 27. $41-8+10=43$

28. $44-9+8=43$ 29. $55-7+9=57$ 30. $53-9+10=54$

Name: Lesson 8-9 Adding and subtracting three numbers

1. $42-4+1=39$ 2. $36-8+1=29$ 3. $61-9+1=53$

4. $35-8+2=29$ 5. $23-6+2=19$ 6. $22-3+2=21$

7. $62-9+3=56$ 8. $65-7+3=61$ 9. $41-6+3=38$

10. $55-9+4=50$ 11. $31-3+4=32$ 12. $57-9+4=52$

13. $41-8+5=38$ 14. $43-4+5=44$ 15. $34-7+5=32$

16. $24-8+6=22$ 17. $58-9+6=55$ 18. $24-5+6=25$

19. $33-6+7=34$ 20. $27-9+7=25$ 21. $51-5+7=53$

22. $63-7+8=64$ 23. $14-9+8=13$ 24. $63-7+8=64$

25. $52-5+9=56$ 26. $52-7+9=54$ 27. $33-6+9=36$

28. $21-4+10=27$ 29. $43-6+10=47$ 30. $56-7+10=59$

Name: Lesson 8-10 Adding and subtracting three numbers

1. $23-7+1=17$ 2. $27-9+1=19$ 3. $42-9+1=34$

4. $46-9+2=39$ 5. $42-4+2=40$ 6. $36-8+2=30$

7. $41-5+3=39$ 8. $35-7+3=31$ 9. $31-6+3=28$

10. $34-7+4=31$ 11. $65-6+4=63$ 12. $22-8+4=18$

13. $53-5+5=53$ 14. $33-4+5=34$ 15. $63-9+5=59$

16. $64-5+6=65$ 17. $56-7+6=55$ 18. $58-9+6=55$

19. $21-9+7=19$ 20. $32-7+7=32$ 21. $31-3+7=35$

22. $65-9+8=64$ 23. $54-5+8=57$ 24. $42-5+8=45$

25. $34-7+9=36$ 26. $32-8+9=33$ 27. $62-8+9=63$

28. $52-8+10=54$ 29. $61-2+10=69$ 30. $55-9+10=56$

Name: **Lesson 9-1 Adding two digit numbers**

1.		54	+ 20	74
2.		35	+ 23	58
3.		36	+ 52	88
4.		86	+ 23	109
5.		72	+ 13	85
6.		27	+ 72	99
7.		25	+ 41	66
8.		32	+ 12	44
9.	1	85	+ 16	101
10.		46	+ 23	69
11.		72	+ 30	102
12.	1	53	+ 28	81
13.	1	24	+ 18	42
14.	1	39	+ 31	70
15.	1	43	+ 39	82
16.	1	35	+ 46	81
17.	1	44	+ 36	80
18.	1	66	+ 28	94
19.	1	27	+ 66	93
20.	1	29	+ 49	78
21.		32	+ 95	127
22.		96	+ 82	178
23.		93	+ 24	117
24.		87	+ 70	157
25.	1	67	+ 53	120
26.		85	+ 44	129
27.		78	+ 61	139
28.		42	+ 72	114
29.		75	+ 82	157
30.		53	+ 55	108

Name: **Lesson 9-2 Adding two digit numbers**

1.		34	+ 61	95
2.		53	+ 30	83
3.		42	+ 44	86
4.		57	+ 12	69
5.		84	+ 13	97
6.	1	48	+ 12	60
7.		36	+ 51	87
8.		73	+ 13	86
9.	1	95	+ 26	121
10.		43	+ 45	88
11.		80	+ 22	102
12.	1	68	+ 13	81
13.	1	38	+ 32	70
14.	1	45	+ 15	60
15.	1	59	+ 23	82
16.	1	57	+ 13	70
17.	1	89	+ 15	104
18.	1	29	+ 43	72
19.	1	68	+ 26	94
20.	1	39	+ 22	61
21.		47	+ 61	108
22.		81	+ 24	105
23.	1	79	+ 51	130
24.		96	+ 62	158
25.		94	+ 75	169
26.		91	+ 70	161
27.		36	+ 93	129
28.		84	+ 51	135
29.	1	87	+ 53	140
30.		65	+ 63	128

Name: **Lesson 9-3 Adding two digit numbers**

1.		54	+ 12	66
2.		26	+ 41	67
3.		28	+ 60	88
4.		32	+ 20	52
5.		67	+ 21	88
6.		24	+ 71	95
7.		35	+ 14	49
8.	1	56	+ 34	90
9.		28	+ 51	79
10.		50	+ 23	73
11.	1	59	+ 31	90
12.	1	27	+ 23	50
13.	1	75	+ 17	92
14.	1	67	+ 17	84
15.	1	38	+ 52	90
16.	1	38	+ 27	65
17.	1	45	+ 48	93
18.	1	58	+ 28	86
19.	1	35	+ 45	80
20.	1	35	+ 16	51
21.		44	+ 74	118
22.		72	+ 32	104
23.		43	+ 65	108
24.		74	+ 71	145
25.		33	+ 76	109
26.		72	+ 83	155
27.		63	+ 52	115
28.		85	+ 44	129
29.		23	+ 84	107
30.		96	+ 80	176

Name: **Lesson 9-4 Adding two digit numbers**

1.		53	+ 41	94
2.		84	+ 11	95
3.	1	33	+ 57	90
4.		94	+ 15	109
5.		65	+ 23	88
6.	1	74	+ 16	90
7.		43	+ 55	98
8.		54	+ 24	78
9.		36	+ 51	87
10.		24	+ 42	66
11.	1	67	+ 23	90
12.	1	37	+ 25	62
13.	1	23	+ 39	62
14.	1	77	+ 14	91
15.	1	69	+ 17	86
16.	1	89	+ 22	111
17.	1	69	+ 31	100
18.	1	46	+ 46	92
19.	1	59	+ 21	80
20.	1	49	+ 18	67
21.		93	+ 42	135
22.		92	+ 31	123
23.		93	+ 73	166
24.	1	27	+ 89	116
25.		45	+ 62	107
26.	1	79	+ 97	176
27.	1	78	+ 62	140
28.		95	+ 33	128
29.		63	+ 56	119
30.	1	73	+ 67	140

Name:

Lesson 9-5 Adding two digit numbers

1. 63 + 26 = 89	2. 56 + 13 = 69	3. 82 + 23 = 105	4. 82 + 32 = 114	5. 47 + 33 = 80
6. 25 + 44 = 69	7. 32 + 27 = 59	8. 53 + 31 = 84	9. 53 + 53 = 106	10. 54 + 21 = 75
11. 36 + 34 = 70	12. 33 + 39 = 72	13. 37 + 43 = 80	14. 37 + 43 = 80	15. 67 + 27 = 94
16. 86 + 18 = 104	17. 67 + 16 = 83	18. 45 + 36 = 81	19. 45 + 36 = 81	20. 69 + 29 = 98
21. 85 + 64 = 149	22. 32 + 75 = 107	23. 85 + 41 = 126	24. 85 + 41 = 126	25. 94 + 13 = 107
26. 69 + 71 = 140	27. 32 + 82 = 114	28. 76 + 81 = 157	29. 76 + 92 = 168	30. 43 + 92 = 135

Name:

Lesson 9-6 Adding two digit numbers

1. 47 + 31 = 78	2. 66 + 24 = 90	3. 35 + 34 = 69	4. 35 + 64 = 99	5. 16 + 61 = 77
6. 43 + 22 = 65	7. 48 + 12 = 60	8. 43 + 15 = 58	9. 93 + 23 = 116	10. 45 + 75 = 120
11. 73 + 39 = 112	12. 46 + 15 = 61	13. 45 + 56 = 101	14. 69 + 17 = 86	15. 34 + 67 = 101
16. 33 + 69 = 102	17. 79 + 28 = 107	18. 68 + 35 = 103	19. 45 + 29 = 74	20. 36 + 24 = 60
21. 72 + 34 = 106	22. 84 + 72 = 156	23. 63 + 92 = 155	24. 28 + 92 = 120	25. 82 + 33 = 115
26. 43 + 72 = 115	27. 87 + 83 = 170	28. 75 + 63 = 138	29. 93 + 95 = 188	30. 45 + 93 = 138

Name:

Lesson 9-7 Adding two digit numbers

1. 75 + 13 = 88	2. 75 + 44 = 119	3. 23 + 74 = 97	4. 54 + 24 = 78	5. 41 + 93 = 134
6. 26 + 31 = 57	7. 84 + 45 = 129	8. 26 + 44 = 70	9. 23 + 49 = 72	10. 32 + 48 = 80
11. 35 + 15 = 50	12. 30 + 29 = 59	13. 39 + 29 = 68	14. 45 + 65 = 110	15. 69 + 96 = 165
16. 38 + 16 = 54	17. 59 + 24 = 83	18. 79 + 34 = 113	19. 69 + 23 = 92	20. 28 + 75 = 103
21. 24 + 93 = 117	22. 94 + 47 = 141	23. 33 + 97 = 130	24. 99 + 99 = 198	25. 64 + 44 = 108
26. 83 + 85 = 168	27. 69 + 62 = 131	28. 68 + 43 = 111	29. 78 + 87 = 165	30. 48 + 49 = 97

Name:

Lesson 9-8 Adding two digit numbers

1. 34 + 24 = 58	2. 25 + 15 = 40	3. 46 + 14 = 60	4. 33 + 77 = 110	5. 54 + 25 = 79
6. 43 + 67 = 110	7. 84 + 23 = 107	8. 62 + 64 = 126	9. 27 + 36 = 63	10. 39 + 42 = 81
11. 46 + 24 = 70	12. 58 + 32 = 90	13. 28 + 76 = 104	14. 63 + 29 = 92	15. 65 + 25 = 90
16. 47 + 26 = 73	17. 79 + 38 = 117	18. 34 + 38 = 72	19. 28 + 65 = 93	20. 83 + 27 = 110
21. 63 + 97 = 160	22. 53 + 82 = 135	23. 57 + 52 = 109	24. 45 + 84 = 129	25. 91 + 19 = 110
26. 93 + 72 = 165	27. 99 + 32 = 131	28. 47 + 73 = 120	29. 92 + 82 = 174	30. 44 + 76 = 120

Name: **Lesson 9-9 Adding two digit numbers**

1. 34 + 36 = 70	2. 62 + 36 = 98	3. 23 + 44 = 67	4. 73 + 37 = 110	5. 41 + 72 = 113
6. 56 + 42 = 98	7. 14 + 76 = 90	8. 55 + 46 = 101	9. 82 + 29 = 111	10. 22 + 83 = 105
11. 78 + 12 = 90	12. 23 + 99 = 122	13. 29 + 38 = 67	14. 45 + 27 = 72	15. 25 + 75 = 100
16. 47 + 33 = 80	17. 36 + 68 = 104	18. 47 + 37 = 84	19. 48 + 84 = 132	20. 43 + 39 = 82
21. 59 + 51 = 110	22. 82 + 62 = 144	23. 43 + 98 = 141	24. 36 + 77 = 113	25. 91 + 93 = 184
26. 33 + 77 = 110	27. 63 + 77 = 140	28. 96 + 95 = 191	29. 35 + 75 = 110	30. 36 + 97 = 133

Name: **Lesson 9-10 Adding two digit numbers**

1. 24 + 46 = 70	2. 76 + 13 = 89	3. 47 + 27 = 74	4. 54 + 54 = 108	5. 84 + 25 = 109
6. 35 + 41 = 76	7. 67 + 26 = 93	8. 25 + 26 = 51	9. 62 + 23 = 85	10. 41 + 92 = 133
11. 57 + 75 = 132	12. 39 + 31 = 70	13. 49 + 28 = 77	14. 36 + 24 = 60	15. 59 + 33 = 92
16. 49 + 49 = 98	17. 33 + 78 = 111	18. 38 + 55 = 93	19. 52 + 58 = 110	20. 78 + 17 = 95
21. 25 + 98 = 123	22. 42 + 29 = 71	23. 23 + 97 = 120	24. 63 + 69 = 132	25. 53 + 46 = 99
26. 78 + 87 = 165	27. 75 + 77 = 152	28. 73 + 38 = 111	29. 95 + 96 = 191	30. 85 + 56 = 141

Name: **Lesson 10-1 Subtracting two digit numbers**

1. 72 − 14 = 58	2. 53 − 27 = 26	3. 84 − 36 = 48	4. 96 − 58 = 38	5. 44 − 19 = 25
6. 96 − 68 = 28	7. 62 − 24 = 38	8. 98 − 39 = 59	9. 72 − 39 = 33	10. 59 − 26 = 33
11. 91 − 13 = 78	12. 85 − 31 = 54	13. 56 − 28 = 28	14. 97 − 58 = 39	15. 73 − 52 = 21
16. 85 − 36 = 49	17. 81 − 68 = 13	18. 53 − 16 = 37	19. 74 − 25 = 49	20. 89 − 38 = 51
21. 82 − 37 = 45	22. 75 − 44 = 31	23. 62 − 34 = 28	24. 81 − 66 = 15	25. 53 − 35 = 18
26. 31 − 17 = 14	27. 84 − 56 = 28	28. 74 − 49 = 25	29. 68 − 29 = 39	30. 62 − 36 = 26

Name: **Lesson 10-2 Subtracting two digit numbers**

1. 95 − 46 = 49	2. 82 − 32 = 50	3. 93 − 13 = 80	4. 52 − 25 = 27	5. 61 − 17 = 44
6. 88 − 59 = 29	7. 83 − 15 = 68	8. 98 − 79 = 19	9. 86 − 19 = 67	10. 84 − 49 = 35
11. 93 − 17 = 76	12. 94 − 47 = 47	13. 77 − 29 = 48	14. 99 − 24 = 75	15. 55 − 37 = 18
16. 45 − 28 = 17	17. 75 − 28 = 47	18. 47 − 18 = 29	19. 42 − 17 = 25	20. 91 − 48 = 43
21. 73 − 55 = 18	22. 65 − 49 = 16	23. 76 − 38 = 38	24. 53 − 27 = 26	25. 45 − 17 = 28
26. 95 − 56 = 39	27. 72 − 58 = 14	28. 52 − 16 = 36	29. 65 − 38 = 27	30. 81 − 15 = 66

Name:

Lesson 10-3 Subtracting two digit numbers

1. 82 − 17 = 65	2. 71 − 33 = 38	3. 62 − 24 = 38	4. 54 − 28 = 26	5. 83 − 45 = 38
6. 83 − 65 = 18	7. 77 − 49 = 28	8. 82 − 35 = 47	9. 73 − 27 = 46	10. 44 − 17 = 27
11. 57 − 18 = 39	12. 73 − 36 = 37	13. 84 − 47 = 37	14. 71 − 18 = 53	15. 82 − 23 = 59
16. 51 − 19 = 32	17. 95 − 58 = 37	18. 31 − 15 = 16	19. 43 − 26 = 17	20. 64 − 39 = 25
21. 71 − 27 = 44	22. 63 − 34 = 29	23. 65 − 27 = 38	24. 66 − 38 = 28	25. 42 − 18 = 24
26. 54 − 16 = 38	27. 46 − 17 = 29	28. 87 − 49 = 38	29. 52 − 37 = 15	30. 92 − 57 = 35

Name:

Lesson 10-4 Subtracting two digit numbers

1. 71 − 58 = 13	2. 83 − 36 = 47	3. 81 − 44 = 37	4. 93 − 57 = 36	5. 61 − 22 = 39
6. 53 − 19 = 34	7. 95 − 47 = 48	8. 52 − 15 = 37	9. 97 − 65 = 32	10. 55 − 38 = 17
11. 64 − 37 = 27	12. 88 − 59 = 29	13. 85 − 49 = 36	14. 47 − 19 = 28	15. 31 − 13 = 18
16. 74 − 48 = 26	17. 62 − 34 = 28	18. 97 − 79 = 18	19. 65 − 15 = 50	20. 48 − 14 = 34
21. 72 − 59 = 13	22. 61 − 23 = 38	23. 77 − 48 = 29	24. 82 − 51 = 31	25. 84 − 38 = 46
26. 85 − 38 = 47	27. 72 − 39 = 33	28. 73 − 16 = 57	29. 82 − 44 = 38	30. 57 − 28 = 29

Name:

Lesson 10-5 Subtracting two digit numbers

1. 62 − 23 = 39	2. 81 − 46 = 35	3. 55 − 29 = 26	4. 45 − 17 = 28	5. 83 − 29 = 54
6. 54 − 19 = 35	7. 83 − 24 = 59	8. 67 − 48 = 19	9. 33 − 16 = 17	10. 85 − 58 = 27
11. 46 − 14 = 32	12. 73 − 29 = 44	13. 65 − 26 = 39	14. 52 − 26 = 26	15. 75 − 37 = 38
16. 53 − 25 = 28	17. 71 − 38 = 33	18. 73 − 46 = 27	19. 86 − 27 = 59	20. 88 − 49 = 39
21. 73 − 36 = 37	22. 55 − 26 = 29	23. 82 − 48 = 34	24. 71 − 34 = 37	25. 63 − 28 = 35
26. 95 − 57 = 38	27. 56 − 17 = 39	28. 64 − 19 = 45	29. 61 − 25 = 36	30. 62 − 34 = 28

Name:

Lesson 10-6 Subtracting two digit numbers

1. 85 − 56 = 29	2. 66 − 29 = 37	3. 92 − 55 = 37	4. 81 − 14 = 67	5. 55 − 19 = 36
6. 62 − 16 = 46	7. 92 − 37 = 55	8. 84 − 48 = 36	9. 75 − 28 = 47	10. 94 − 57 = 37
11. 91 − 42 = 49	12. 64 − 49 = 15	13. 33 − 16 = 17	14. 96 − 68 = 28	15. 42 − 16 = 26
16. 66 − 39 = 27	17. 82 − 55 = 27	18. 45 − 26 = 19	19. 81 − 47 = 34	20. 73 − 19 = 54
21. 84 − 65 = 19	22. 54 − 29 = 25	23. 72 − 38 = 34	24. 45 − 19 = 26	25. 97 − 69 = 28
26. 86 − 18 = 68	27. 93 − 79 = 14	28. 92 − 54 = 38	29. 64 − 27 = 37	30. 76 − 48 = 28

Name: **Lesson 10-7 Subtracting two digit numbers**

1. 76 − 59 = 17	2. 52 − 18 = 34	3. 51 − 26 = 25	4. 68 − 19 = 49	5. 84 − 37 = 47
6. 73 − 37 = 36	7. 95 − 19 = 76	8. 97 − 39 = 58	9. 42 − 16 = 26	10. 66 − 28 = 38
11. 84 − 48 = 36	12. 77 − 39 = 38	13. 63 − 36 = 27	14. 54 − 29 = 25	15. 63 − 19 = 44
16. 94 − 17 = 77	17. 63 − 25 = 38	18. 33 − 18 = 15	19. 91 − 52 = 39	20. 55 − 28 = 27
21. 73 − 49 = 24	22. 61 − 33 = 28	23. 94 − 39 = 55	24. 97 − 68 = 29	25. 91 − 36 = 55
26. 45 − 18 = 27	27. 92 − 35 = 57	28. 86 − 48 = 38	29. 95 − 66 = 29	30. 86 − 29 = 57

Name: **Lesson 10-8 Subtracting two digit numbers**

1. 85 − 66 = 19	2. 53 − 18 = 35	3. 81 − 55 = 26	4. 62 − 27 = 35	5. 92 − 39 = 53
6. 73 − 15 = 58	7. 89 − 49 = 40	8. 93 − 16 = 77	9. 75 − 39 = 36	10. 72 − 44 = 28
11. 82 − 28 = 54	12. 75 − 68 = 7	13. 61 − 15 = 46	14. 94 − 36 = 58	15. 82 − 33 = 49
16. 95 − 39 = 56	17. 93 − 68 = 25	18. 72 − 16 = 56	19. 81 − 14 = 67	20. 94 − 69 = 25
21. 71 − 46 = 25	22. 92 − 25 = 67	23. 55 − 18 = 37	24. 86 − 59 = 27	25. 62 − 17 = 45
26. 55 − 27 = 28	27. 87 − 58 = 29	28. 67 − 29 = 38	29. 92 − 28 = 64	30. 93 − 69 = 24

Name: **Lesson 10-9 Subtracting two digit numbers**

1. 72 − 25 = 47	2. 85 − 48 = 37	3. 93 − 29 = 64	4. 93 − 41 = 52	5. 43 − 26 = 17
6. 55 − 39 = 16	7. 63 − 15 = 48	8. 97 − 39 = 58	9. 92 − 57 = 35	10. 72 − 33 = 39
11. 65 − 28 = 37	12. 53 − 19 = 34	13. 84 − 55 = 29	14. 81 − 58 = 23	15. 62 − 44 = 18
16. 92 − 44 = 48	17. 98 − 38 = 60	18. 85 − 28 = 57	19. 56 − 17 = 39	20. 94 − 59 = 35
21. 95 − 16 = 79	22. 95 − 27 = 68	23. 73 − 49 = 24	24. 71 − 34 = 37	25. 83 − 39 = 44
26. 83 − 17 = 66	27. 85 − 39 = 46	28. 64 − 27 = 37	29. 91 − 46 = 45	30. 93 − 69 = 24

Name: **Lesson 10-10 Subtracting two digit numbers**

1. 65 − 26 = 39	2. 72 − 27 = 45	3. 92 − 28 = 64	4. 83 − 27 = 56	5. 95 − 68 = 27
6. 73 − 35 = 38	7. 75 − 49 = 26	8. 95 − 39 = 56	9. 65 − 36 = 29	10. 58 − 19 = 39
11. 93 − 28 = 65	12. 63 − 25 = 38	13. 64 − 25 = 39	14. 57 − 19 = 38	15. 74 − 58 = 16
16. 92 − 59 = 33	17. 55 − 29 = 26	18. 65 − 39 = 26	19. 86 − 37 = 49	20. 91 − 66 = 25
21. 65 − 27 = 38	22. 51 − 28 = 23	23. 51 − 27 = 24	24. 52 − 14 = 38	25. 93 − 48 = 45
26. 46 − 19 = 27	27. 62 − 47 = 15	28. 82 − 46 = 36	29. 71 − 33 = 38	30. 55 − 27 = 28

Name: Lesson 11-1 Adding three numbers

1. $12 + 15 + 1 = 28$	2. $14 + 14 + 2 = 30$	3. $12 + 19 + 3 = 34$
4. $11 + 12 + 1 = 24$	5. $12 + 16 + 2 = 30$	6. $18 + 14 + 3 = 35$
7. $13 + 14 + 1 = 28$	8. $16 + 16 + 2 = 34$	9. $13 + 12 + 3 = 28$
10. $17 + 11 + 1 = 29$	11. $13 + 14 + 2 = 29$	12. $15 + 15 + 3 = 33$
13. $15 + 13 + 1 = 29$	14. $13 + 17 + 2 = 32$	15. $14 + 17 + 3 = 34$
16. $11 + 16 + 1 = 28$	17. $15 + 17 + 2 = 34$	18. $19 + 19 + 3 = 41$
19. $14 + 17 + 1 = 32$	20. $14 + 18 + 2 = 34$	21. $14 + 15 + 3 = 32$
22. $18 + 16 + 1 = 35$	23. $19 + 17 + 2 = 38$	24. $12 + 14 + 3 = 29$
25. $12 + 19 + 1 = 32$	26. $18 + 13 + 2 = 33$	27. $12 + 19 + 3 = 34$
28. $13 + 17 + 1 = 31$	29. $16 + 15 + 2 = 33$	30. $15 + 17 + 3 = 35$

Name: Lesson 11-2 Adding three numbers

1. $11 + 15 + 2 = 28$	2. $17 + 13 + 3 = 33$	3. $12 + 18 + 4 = 34$
4. $13 + 15 + 2 = 30$	5. $18 + 19 + 3 = 40$	6. $13 + 17 + 4 = 34$
7. $18 + 17 + 2 = 37$	8. $13 + 15 + 3 = 31$	9. $14 + 17 + 4 = 35$
10. $14 + 15 + 2 = 31$	11. $18 + 19 + 3 = 40$	12. $16 + 17 + 4 = 37$
13. $15 + 19 + 2 = 36$	14. $12 + 17 + 3 = 32$	15. $13 + 16 + 4 = 33$
16. $16 + 14 + 2 = 32$	17. $18 + 13 + 3 = 34$	18. $15 + 19 + 4 = 38$
19. $13 + 17 + 2 = 32$	20. $12 + 16 + 3 = 31$	21. $16 + 18 + 4 = 38$
22. $12 + 18 + 2 = 32$	23. $15 + 15 + 3 = 33$	24. $11 + 19 + 4 = 34$
25. $17 + 17 + 2 = 36$	26. $19 + 15 + 3 = 37$	27. $14 + 16 + 4 = 34$
28. $16 + 15 + 2 = 33$	29. $16 + 16 + 3 = 35$	30. $19 + 19 + 4 = 42$

Name: Lesson 11-3 Adding three numbers

1. $16 + 11 + 3 = 30$	2. $14 + 14 + 4 = 32$	3. $11 + 19 + 5 = 35$
4. $12 + 14 + 3 = 29$	5. $17 + 11 + 4 = 32$	6. $13 + 14 + 5 = 32$
7. $11 + 13 + 3 = 27$	8. $12 + 15 + 4 = 31$	9. $11 + 12 + 5 = 38$
10. $13 + 12 + 3 = 28$	11. $14 + 13 + 4 = 31$	12. $14 + 11 + 5 = 30$
13. $11 + 15 + 3 = 29$	14. $17 + 12 + 4 = 33$	15. $15 + 14 + 5 = 34$
16. $13 + 16 + 3 = 32$	17. $14 + 12 + 4 = 30$	18. $12 + 13 + 5 = 30$
19. $17 + 11 + 3 = 31$	20. $16 + 12 + 4 = 32$	21. $12 + 17 + 5 = 34$
22. $13 + 15 + 3 = 31$	23. $11 + 14 + 4 = 29$	24. $11 + 16 + 5 = 32$
25. $12 + 11 + 3 = 26$	26. $15 + 13 + 4 = 32$	27. $13 + 13 + 5 = 31$
28. $15 + 19 + 3 = 37$	29. $18 + 11 + 4 = 33$	30. $14 + 15 + 5 = 34$

Name: Lesson 11-4 Adding three numbers

1. $13 + 13 + 4 = 30$	2. $16 + 11 + 5 = 32$	3. $12 + 14 + 6 = 32$
4. $12 + 11 + 4 = 27$	5. $11 + 15 + 5 = 31$	6. $14 + 11 + 6 = 31$
7. $18 + 11 + 4 = 33$	8. $12 + 13 + 5 = 30$	9. $11 + 18 + 6 = 35$
10. $16 + 13 + 4 = 33$	11. $14 + 15 + 5 = 34$	12. $13 + 14 + 6 = 33$
13. $14 + 13 + 4 = 31$	14. $12 + 16 + 5 = 33$	15. $15 + 16 + 6 = 37$
16. $11 + 17 + 4 = 32$	17. $17 + 12 + 5 = 34$	18. $15 + 19 + 6 = 40$
19. $11 + 14 + 4 = 29$	20. $13 + 14 + 5 = 32$	21. $14 + 14 + 6 = 34$
22. $16 + 12 + 4 = 32$	23. $12 + 18 + 5 = 35$	24. $13 + 16 + 6 = 35$
25. $15 + 14 + 4 = 33$	26. $11 + 19 + 5 = 35$	27. $17 + 18 + 6 = 41$
28. $13 + 15 + 4 = 32$	29. $15 + 16 + 5 = 36$	30. $14 + 19 + 6 = 39$

Name: Lesson 11-5 Adding three numbers

1. $18+11+5=34$ 2. $15+12+6=33$ 3. $14+11+7=32$

4. $14+12+5=31$ 5. $13+16+6=35$ 6. $12+17+7=36$

7. $11+15+5=31$ 8. $11+14+6=31$ 9. $11+17+7=35$

10. $14+13+5=32$ 11. $13+11+6=30$ 12. $12+16+7=35$

13. $11+17+5=33$ 14. $12+13+6=31$ 15. $15+15+7=37$

16. $12+15+5=32$ 17. $13+15+6=34$ 18. $16+13+7=36$

19. $13+13+5=31$ 20. $12+18+6=36$ 21. $14+12+7=33$

22. $14+15+5=34$ 23. $13+18+6=37$ 24. $15+16+7=38$

25. $17+13+5=35$ 26. $14+14+6=34$ 27. $17+18+7=42$

28. $16+17+5=38$ 29. $15+17+6=38$ 30. $16+19+7=42$

Name: Lesson 11-6 Adding three numbers

1. $13+12+6=31$ 2. $11+18+7=36$ 3. $13+11+8=32$

4. $15+11+6=32$ 5. $12+12+7=31$ 6. $11+16+8=35$

7. $14+13+6=33$ 8. $16+12+7=35$ 9. $13+13+8=34$

10. $12+15+6=33$ 11. $17+11+7=35$ 12. $14+11+8=33$

13. $15+14+6=35$ 14. $15+12+7=34$ 15. $13+16+8=37$

16. $14+14+6=34$ 17. $12+14+7=33$ 18. $13+17+8=38$

19. $17+12+6=35$ 20. $15+13+7=35$ 21. $18+11+8=37$

22. $16+13+6=35$ 23. $11+15+7=33$ 24. $12+17+8=37$

25. $11+14+6=31$ 26. $12+16+7=35$ 27. $14+15+8=40$

28. $17+15+6=38$ 29. $13+14+7=34$ 30. $14+19+8=41$

Name: Lesson 11-7 Adding three numbers

1. $18+11+7=36$ 2. $14+13+8=35$ 3. $11+12+9=32$

4. $11+15+7=33$ 5. $17+11+8=36$ 6. $13+12+9=34$

7. $12+14+7=33$ 8. $11+12+8=31$ 9. $15+12+9=36$

10. $16+13+7=36$ 11. $16+11+8=35$ 12. $14+15+9=38$

13. $15+12+7=34$ 14. $11+14+8=33$ 15. $17+12+9=38$

16. $14+13+7=34$ 17. $12+17+8=37$ 18. $13+13+9=35$

19. $12+19+7=38$ 20. $14+14+8=36$ 21. $12+16+9=37$

22. $14+11+7=32$ 23. $12+19+8=39$ 24. $11+17+9=37$

25. $15+14+7=36$ 26. $13+15+8=36$ 27. $19+12+9=40$

28. $18+17+7=42$ 29. $12+18+8=38$ 30. $13+17+9=39$

Name: Lesson 11-8 Adding three numbers

1. $14+11+8=33$ 2. $16+13+9=38$ 3. $11+15+10=36$

4. $15+14+8=37$ 5. $12+14+9=41$ 6. $13+14+10=37$

7. $16+11+8=35$ 8. $15+12+9=36$ 9. $14+14+10=38$

10. $13+12+8=33$ 11. $11+13+9=33$ 12. $18+11+10=39$

13. $17+11+8=36$ 14. $13+14+9=36$ 15. $13+16+10=39$

16. $12+16+8=36$ 17. $12+17+9=38$ 18. $11+14+10=35$

19. $15+13+8=36$ 20. $12+13+9=34$ 21. $16+12+10=38$

22. $14+13+8=35$ 23. $15+17+9=41$ 24. $14+15+10=39$

25. $12+15+8=35$ 26. $18+12+9=39$ 27. $15+11+10=36$

28. $15+16+8=39$ 29. $17+19+9=45$ 30. $11+17+10=38$

Name: **Lesson 11-9 Adding three numbers**

1. $12+14+1=27$
2. $14+15+1=30$
3. $12+17+1=30$
4. $15+12+2=29$
5. $17+11+2=30$
6. $13+12+2=27$
7. $13+15+3=31$
8. $18+11+3=32$
9. $12+15+3=30$
10. $14+14+4=32$
11. $12+16+4=32$
12. $11+12+4=28$
13. $12+13+5=30$
14. $11+14+5=30$
15. $11+17+5=33$
16. $16+11+6=33$
17. $13+13+6=32$
18. $13+16+6=35$
19. $11+15+7=33$
20. $14+13+7=34$
21. $15+11+7=33$
22. $16+12+8=36$
23. $11+18+8=37$
24. $11+16+8=35$
25. $14+12+9=35$
26. $17+12+9=38$
27. $14+11+9=34$
28. $15+14+10=39$
29. $16+18+10=44$
30. $13+19+10=42$

Name: **Lesson 11-10 Adding three numbers**

1. $13+12+1=26$
2. $16+13+1=30$
3. $12+12+1=25$
4. $15+14+2=31$
5. $12+15+2=29$
6. $14+13+2=29$
7. $17+11+3=31$
8. $14+12+3=29$
9. $16+11+3=30$
10. $15+13+4=32$
11. $11+12+4=27$
12. $17+13+4=34$
13. $11+15+5=31$
14. $11+17+5=33$
15. $11+18+5=34$
16. $14+11+6=31$
17. $17+13+6=36$
18. $13+16+6=35$
19. $18+11+7=36$
20. $15+16+7=38$
21. $15+15+7=37$
22. $13+16+8=37$
23. $16+16+8=40$
24. $14+19+8=41$
25. $13+19+9=41$
26. $14+18+9=41$
27. $13+18+9=40$
28. $17+18+10=45$
29. $17+19+10=46$
30. $18+19+10=47$

Name: **Lesson 12-1 Adding and subtracting three numbers**

1. $24-11+1=14$
2. $22-16+2=8$
3. $28-19+3=11$
4. $28-15+1=14$
5. $24-15+2=11$
6. $26-19+3=10$
7. $26-14+1=13$
8. $22-13+2=11$
9. $23-14+3=12$
10. $28-17+1=12$
11. $27-19+2=10$
12. $26-18+3=11$
13. $22-15+1=8$
14. $24-19+2=7$
15. $21-15+3=9$
16. $25-17+1=9$
17. $25-18+2=9$
18. $16-13+3=6$
19. $23-9+1=15$
20. $24-17+2=9$
21. $27-18+3=12$
22. $23-9+1=15$
23. $21-18+2=5$
24. $23-12+3=14$
25. $22-8+1=15$
26. $25-19+2=8$
27. $21-13+3=11$
28. $25-9+1=17$
29. $23-17+2=8$
30. $25-19+3=9$

Name: **Lesson 12-2 Adding and subtracting three numbers**

1. $38-12+2=28$
2. $33-16+3=20$
3. $33-19+4=18$
4. $32-15+2=19$
5. $31-14+3=20$
6. $36-17+4=23$
7. $37-19+2=20$
8. $35-16+3=22$
9. $38-19+4=23$
10. $33-17+2=18$
11. $32-13+3=22$
12. $34-18+4=20$
13. $36-18+2=20$
14. $34-17+3=20$
15. $34-15+4=23$
16. $33-14+2=21$
17. $32-14+3=21$
18. $36-19+4=21$
19. $31-16+2=17$
20. $31-18+3=16$
21. $31-12+4=23$
22. $32-19+2=15$
23. $34-16+3=21$
24. $31-17+4=18$
25. $31-15+2=18$
26. $35-18+3=20$
27. $35-19+4=20$
28. $34-19+2=17$
29. $32-16+3=19$
30. $37-19+4=22$

Name: Lesson 12-3 Adding and subtracting three numbers

1. $33-13+3=23$ 2. $38-19+4=23$ 3. $34-15+5=24$

4. $33-18+3=18$ 5. $37-19+4=22$ 6. $31-14+5=22$

7. $31-16+3=18$ 8. $35-17+4=22$ 9. $33-16+5=22$

10. $35-19+3=19$ 11. $37-18+4=23$ 12. $33-19+5=19$

13. $33-17+3=19$ 14. $31-15+4=20$ 15. $36-18+5=23$

16. $31-12+3=22$ 17. $34-16+4=22$ 18. $32-19+5=18$

19. $34-18+3=19$ 20. $31-13+4=22$ 21. $31-17+5=19$

22. $33-14+3=22$ 23. $34-17+4=21$ 24. $36-19+5=22$

25. $32-16+3=19$ 26. $32-18+4=18$ 27. $35-16+5=24$

28. $31-19+3=16$ 29. $32-13+4=23$ 30. $35-18+5=22$

Name: Lesson 12-4 Adding and subtracting three numbers

1. $42-12+4=34$ 2. $44-16+5=33$ 3. $43-17+6=32$

4. $43-14+4=33$ 5. $43-19+5=29$ 6. $41-15+6=32$

7. $41-12+4=33$ 8. $44-18+5=31$ 9. $42-14+6=34$

10. $43-15+4=32$ 11. $48-19+5=34$ 12. $41-13+6=34$

13. $46-17+4=33$ 14. $41-14+5=32$ 15. $45-18+6=33$

16. $44-17+4=31$ 17. $42-16+5=31$ 18. $43-19+6=30$

19. $41-18+4=27$ 20. $43-18+5=30$ 21. $45-16+6=35$

22. $47-19+4=32$ 23. $43-16+5=32$ 24. $46-18+6=34$

25. $42-13+4=33$ 26. $41-19+5=27$ 27. $45-17+6=34$

28. $45-19+4=30$ 29. $42-15+5=32$ 30. $46-18+6=34$

Name: Lesson 12-5 Adding and subtracting three numbers

1. $46-16+5=35$ 2. $44-15+6=35$ 3. $44-15+7=36$

4. $45-17+5=33$ 5. $47-19+6=34$ 6. $47-19+7=35$

7. $41-13+5=33$ 8. $42-17+6=31$ 9. $42-17+7=32$

10. $45-18+5=32$ 11. $44-19+6=31$ 12. $44-19+7=32$

13. $44-17+5=32$ 14. $41-16+6=31$ 15. $41-16+7=32$

16. $43-16+5=32$ 17. $43-19+6=30$ 18. $43-19+7=31$

19. $42-15+5=32$ 20. $44-16+6=34$ 21. $44-16+7=35$

22. $42-19+5=28$ 23. $42-13+6=35$ 24. $42-13+7=36$

25. $41-18+5=28$ 26. $43-15+6=34$ 27. $42-18+7=31$

28. $45-17+5=33$ 29. $41-19+6=28$ 30. $43-15+7=35$

Name: Lesson 12-6 Adding and subtracting three numbers

1. $43-18+6=31$ 2. $42-14+7=35$ 3. $47-18+8=37$

4. $44-19+6=31$ 5. $41-15+7=33$ 6. $45-17+8=36$

7. $41-12+6=35$ 8. $43-17+7=33$ 9. $43-19+8=32$

10. $43-15+6=34$ 11. $44-15+7=36$ 12. $41-13+8=36$

13. $46-17+6=35$ 14. $41-14+7=34$ 15. $45-16+8=37$

16. $46-18+6=34$ 17. $44-16+7=35$ 18. $45-19+8=34$

19. $46-19+6=33$ 20. $44-18+7=33$ 21. $43-14+8=37$

22. $41-17+6=30$ 23. $42-15+7=34$ 24. $45-18+8=35$

25. $42-16+6=32$ 26. $44-17+7=34$ 27. $41-18+8=41$

28. $43-16+6=33$ 29. $42-13+7=36$ 30. $43-16+8=35$

Name:

Lesson 12-7 Adding and subtracting three numbers

1. $53 - 23 + 7 = 37$ 2. $53 - 16 + 8 = 45$ 3. $53 - 17 + 9 = 45$

4. $51 - 17 + 7 = 41$ 5. $56 - 29 + 8 = 35$ 6. $52 - 25 + 9 = 36$

7. $57 - 19 + 7 = 45$ 8. $57 - 18 + 8 = 47$ 9. $51 - 24 + 9 = 36$

10. $54 - 15 + 7 = 46$ 11. $56 - 28 + 8 = 36$ 12. $54 - 16 + 9 = 47$

13. $52 - 27 + 7 = 32$ 14. $52 - 13 + 8 = 47$ 15. $52 - 22 + 9 = 39$

16. $52 - 16 + 7 = 43$ 17. $55 - 27 + 8 = 36$ 18. $53 - 29 + 9 = 33$

19. $51 - 25 + 7 = 33$ 20. $53 - 18 + 8 = 43$ 21. $58 - 19 + 9 = 48$

22. $54 - 29 + 7 = 32$ 23. $55 - 17 + 8 = 46$ 24. $51 - 16 + 9 = 44$

25. $52 - 18 + 7 = 41$ 26. $51 - 27 + 8 = 32$ 27. $53 - 25 + 9 = 37$

28. $51 - 23 + 7 = 35$ 29. $54 - 15 + 8 = 47$ 30. $54 - 18 + 9 = 45$

Name:

Lesson 12-8 Adding and subtracting three numbers

1. $53 - 26 + 8 = 35$ 2. $54 - 25 + 9 = 38$ 3. $53 - 17 + 10 = 46$

4. $55 - 18 + 8 = 45$ 5. $51 - 12 + 9 = 48$ 6. $51 - 29 + 10 = 32$

7. $51 - 17 + 8 = 42$ 8. $56 - 29 + 9 = 36$ 9. $52 - 23 + 10 = 39$

10. $55 - 26 + 8 = 37$ 11. $52 - 25 + 9 = 36$ 12. $57 - 28 + 10 = 39$

13. $53 - 19 + 8 = 42$ 14. $54 - 16 + 9 = 47$ 15. $56 - 27 + 10 = 39$

16. $55 - 17 + 8 = 46$ 17. $52 - 27 + 9 = 34$ 18. $58 - 19 + 10 = 49$

19. $55 - 29 + 8 = 34$ 20. $51 - 13 + 9 = 47$ 21. $54 - 18 + 10 = 46$

22. $51 - 16 + 8 = 43$ 23. $52 - 28 + 9 = 33$ 24. $53 - 23 + 10 = 40$

25. $52 - 28 + 8 = 32$ 26. $54 - 18 + 9 = 45$ 27. $55 - 27 + 10 = 38$

28. $54 - 17 + 8 = 45$ 29. $51 - 15 + 9 = 45$ 30. $56 - 18 + 10 = 48$

Name:

Lesson 12-9 Adding and subtracting three numbers

1. $45 - 17 + 1 = 29$ 2. $53 - 25 + 1 = 29$ 3. $52 - 29 + 1 = 24$

4. $51 - 23 + 2 = 30$ 5. $45 - 29 + 2 = 18$ 6. $47 - 18 + 2 = 31$

7. $56 - 29 + 3 = 30$ 8. $56 - 28 + 3 = 31$ 9. $51 - 12 + 3 = 42$

10. $43 - 18 + 4 = 29$ 11. $41 - 17 + 4 = 28$ 12. $43 - 26 + 4 = 21$

13. $53 - 19 + 5 = 39$ 14. $45 - 26 + 5 = 24$ 15. $52 - 24 + 5 = 33$

16. $51 - 14 + 6 = 43$ 17. $54 - 18 + 6 = 42$ 18. $41 - 25 + 6 = 22$

19. $44 - 29 + 7 = 22$ 20. $44 - 26 + 7 = 25$ 21. $53 - 17 + 7 = 43$

22. $51 - 19 + 8 = 40$ 23. $52 - 25 + 8 = 35$ 24. $58 - 19 + 8 = 47$

25. $42 - 23 + 9 = 28$ 26. $52 - 16 + 9 = 45$ 27. $47 - 28 + 9 = 28$

28. $52 - 18 + 10 = 44$ 29. $45 - 28 + 10 = 27$ 30. $56 - 17 + 10 = 49$

Name:

Lesson 12-10 Adding and subtracting three numbers

1. $54 - 16 + 1 = 39$ 2. $41 - 29 + 1 = 13$ 3. $51 - 19 + 1 = 33$

4. $55 - 19 + 2 = 38$ 5. $53 - 17 + 2 = 38$ 6. $42 - 19 + 2 = 25$

7. $47 - 28 + 3 = 22$ 8. $54 - 15 + 3 = 42$ 9. $43 - 24 + 3 = 22$

10. $55 - 27 + 4 = 32$ 11. $46 - 18 + 4 = 32$ 12. $52 - 15 + 4 = 41$

13. $41 - 13 + 5 = 38$ 14. $53 - 29 + 5 = 29$ 15. $45 - 19 + 5 = 31$

16. $43 - 25 + 6 = 24$ 17. $45 - 26 + 6 = 25$ 18. $52 - 27 + 6 = 31$

19. $52 - 17 + 7 = 42$ 20. $52 - 14 + 7 = 45$ 21. $43 - 17 + 7 = 33$

22. $51 - 23 + 8 = 36$ 23. $41 - 22 + 8 = 27$ 24. $51 - 25 + 8 = 34$

25. $42 - 17 + 9 = 34$ 26. $52 - 18 + 9 = 43$ 27. $47 - 19 + 9 = 37$

28. $54 - 26 + 10 = 38$ 29. $42 - 15 + 10 = 37$ 30. $56 - 18 + 10 = 48$

www.ingramcontent.com/pod-product-compliance
Lightning Source LLC
Chambersburg PA
CBHW080830220526
45467CB00008B/2248
* 9 7 8 1 7 1 7 0 4 8 5 5 4 *